In Our Own Image

IN OUR OWN IMAGE

Building an Artificial Person

MAUREEN CAUDILL

New York Oxford
OXFORD UNIVERSITY PRESS
1992

Oxford University Press

Oxford New York Toronto
Delhi Bombay Calcutta Madras Karachi
Kuala Lumpur Singapore Hong Kong Tokyo
Nairobi Dar es Salaam Cape Town
Melbourne Auckland

and associated companies in
Berlin Ibadan

Published by Oxford University Press, Inc.
200 Madison Avenue, New York, New York 10016

Library of Congress Cataloging-in-Publication Data
Caudill, Maureen.
In our own image :
building an artificial person /
Maureen Caudill.
p. cm. Includes bibliographical references and index.
ISBN 0–19–507338–X
1. Robotics. 2. Artificial intelligence. I. Title.
TJ211.C28 1992 629.8″92—dc20 91–42524

1 2 3 4 5 6 7 8 9

Printed in the United States of America
on acid-free paper

This book is dedicated to those visionaries who have the courage
to look into tomorrow and dream impossible dreams,
and especially to the memory of Gene Roddenberry,
who shared his impossible dream with the rest of us
and made us believe it might be possible after all. . . .

Preface

A science fiction story written by Eric Frank Russell tells of the heroism of a spaceship pilot when his ship accidentally ventures too near the sun. The pilot has been the target of prejudicial treatment from his crewmates because of his unusual appearance, and he has tolerated social slurs against "his kind." When disaster strikes, however, the pilot heroically endures blistering heat and blinding sun to steer the spaceship through a daring maneuver that saves the lives of all aboard—except himself. His blackened body and blinded eyes seem all that remain after his noble feat. His fellow crewmembers, saddened (although relieved) by his heroic actions, are repentant and only then realize how shabbily they have treated him.

Is this a story of martyrdom and self-sacrifice? No. It is a story of hope, for the pilot in question is an android, a robot in the shape of a human. The pilot, named "Jay Score" (which is also the title of the story) because his manufacturer's model number is J-20, undergoes a renovation and is put back to work on other missions with the same crewmembers, who now insist on his return among them. Androids, it seems, are made of sturdier stuff than people.

Science fiction has long been intrigued with the notion of a mechanical device that mimics a person. If you have seen *Star Wars* or watched *Star Trek: The Next Generation* you know that an android is a robotic system constructed to resemble a human being. Commander Data in *Star Trek* and C-3PO in *Star Wars* are examples. The popular R2-D2 of *Star Wars* only marginally qualifies as an android, since it has a shape only vaguely human. Some other famous androids in science fiction have been Isaac Asimov's positronic robots and Robbie the Robot in the movie *Forbidden Planet*.

But how far out is such fiction? Is it really possible to build an android like those we see on the movie screen or read about in books? If not, how far away are we from being able to build them? And how would they really work?

Imagine that a national priority has been declared to build an artificial person—an android. Suppose that this effort has been raised to the same level that putting a man on the moon held in the 1960s.

How would such a program affect us in our daily lives? And how soon could we expect to find artificial people in the streets?

This book investigates the possibility of building an artificial person. The goal is to see how far along we are in understanding the basic science necessary to build an android, and to explore how what we know now, along with what we expect to know in the near future, will impact the daily lives of all of us.

Many readers will be surprised at how far along we already are in some aspects of this project; others will be disappointed to learn how far we have yet to go. However, it is quite true that success in building even small portions of an android will have a significant impact on us and the way we live. Just as the space program has affected our society in innumerable ways, from freeze-dried coffee to the development of new plastics and ceramics, we can expect efforts to develop a working android to have a similar effect. While the national priority alluded to above has not been declared—and we should not expect it to be declared any time in the near future—many scientists throughout the world are hard at work on solving the key technical problems in such a project. Their immediate goal may not be an android, but the long-term effect of their efforts will certainly be to advance the day when we can create a working version of C-3PO or Commander Data.

San Diego M.C.
December, 1991

Contents

In Our Own Image

≥ 1 ≤

The Measure of Mankind

Some, valuing those of their own size or mind,
Still make themselves the measure of mankind:
Fondly we think we honor merit then,
When we but praise ourselves in other men.

Alexander Pope

To build an artificial person, we must understand what it is we are trying to construct. What are the basic characteristics an android must have for us to accept it as reasonably complete? The list below offers a preliminary set of minimal requirements.

1. It must be able to see and interpret what it sees. The visual interpretation must be sufficiently accurate to enable it to react appropriately to the events around it.
2. It should have a more or less human shape. It should have two arms, two legs, an identifiable head, and a reasonably human-shaped torso. Its size should be between, say, four and ten feet tall. For pragmatic reasons, the limit of two legs may be more or less negotiable. At the very least, its shape should not be too similar to frightening or otherwise unappealing animals such as spiders, snakes, lizards, and so on; such an appearance would certainly make it difficult to be accepted by most humans.
3. It must be able to move itself around. The android must be able to walk through a room of furniture and people without tripping over or stepping on anything or anyone.
4. It must be able to pick up and carry objects of reasonable size with its hands and arms. It must be able to pick up any object of appropriate size that it sees, no matter what the object's relative position and orientation to the android.
5. It must be able to remember important information.

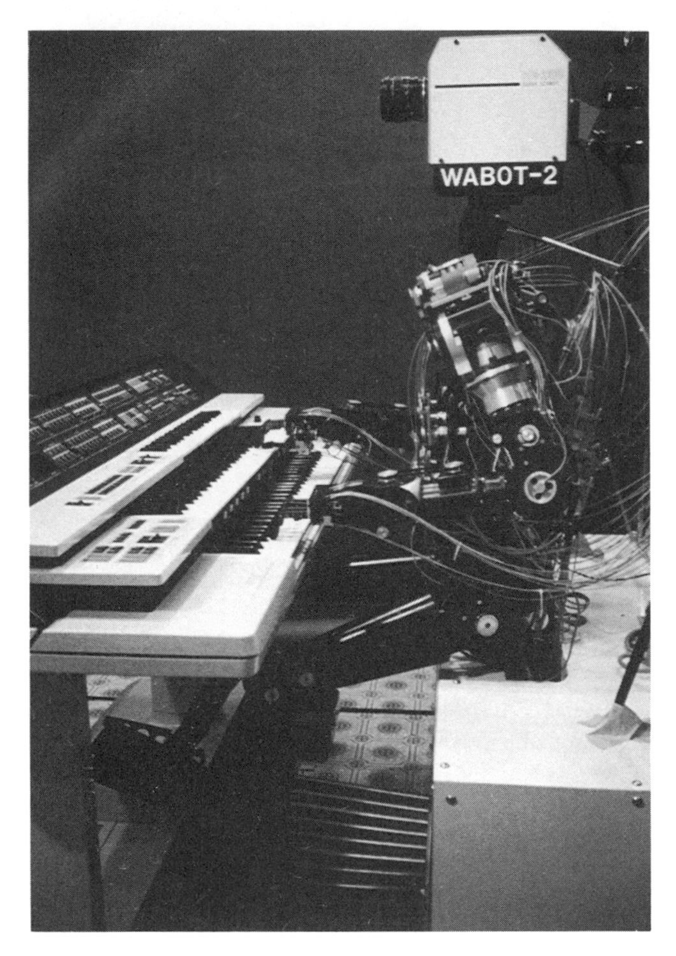

Figure 1.1 WABOT, developed in Japan in the mid-1980s, can read sheet music, play keyboard instruments, accompany a human singer, and carry on simple conversations. It is an early example of an android. (Photo supplied courtesy of Ichiro Kato, KATO-SUGANO Laboratory, Waseda University.)

6. It must be able to accept training so that it can learn to perform the tasks we want it to do.
7. Since we cannot possibly forecast all situations the android will encounter, it must also be able to learn directly from its own experience.
8. It must be able to solve at least simple real-world problems and cope with unexpected situations. If given a general order

("take the trash out" or "do the dishes"), it must be able to figure out how to perform the task appropriately.

9. It must have sensory input devices that can cope with real-world inputs other than vision, such as sounds, touches, and so on. Since these input signals will not be appropriately scaled or formatted for computer use, the android must do its own signal processing to convert the raw data to a usable format.
10. It must be able to communicate with people in English (or some other human language). At a minimum it must be able to comprehend spoken language, and it must be able to generate language either in text form or, preferably, by speech. It should also be able to comprehend other sounds as well, such as warning signals like sirens, bells, and so on.
11. It must have a reasonable amount of common sense, or general knowledge about the world. It should know, for example, that objects fall down if they are not properly supported, that objects can be broken or crushed, and that because something is temporarily obscured does not mean that it has ceased to exist. These are basic principles of the physical world. Furthermore, it should know the difference between beings that are alive, like plants and animals and people, and objects that are not, like buildings and rocks and cars. It should understand that live things need more care in handling than inanimate objects, and should behave appropriately.
12. In addition, the android should understand and obey basic principles of the social world. If an android is to be used for duties such as child care or nursing, it must understand a reasonable amount of human psychology—such as the fact that children sometimes lie—and be able to judge social interactions appropriately. (Are the children playing, or are they really in a serious fight?) It should have good manners and other simple social skills. It should also have a socially acceptable code of behavior—a set of "morals."

The attributes listed here are very basic, and do not consider many physical aspects of the android, such as power source, weight, material composition, and so on; but these mere physical properties need not concern us here. The key aspect of building the android is that it behave properly—if its behavior is acceptable, other physical problems are much easier to solve. Other chapters in this book look at these requirements in detail and try to assess how far away—or how near—we are to meeting them.

You may say that these requirements do not necessarily constitute an "intelligent" device. This is not a trivial issue, and the arguments on each side of this question are both complex and passionate. I would like to set it aside for a moment, however, and return to it only much later in the book; in the meantime, understand that the term "intelligent robot" or "intelligent android" as used in the following chapters refers to any device that meets the requirements listed above.*

In effect, the above list can be considered an operational definition of an intelligent android. Scientists frequently use operational definitions of quantities to ensure precision in meaning. For example, a physicist may specify an operational definition of mass by explaining exactly how it can be measured. Any quantity that results from that measurement technique is *by definition* a mass. For the moment I am creating my own operational definition; any device that meets the criteria I have listed will be considered *by definition* an intelligent android. Later, however, I will come back to the question of whether such a device is truly "intelligent" and consider the arguments—and their implications—on both sides of this issue.

Before beginning the more detailed assessment of current and future android capabilities, however, let's consider how an android that meets these requirements will affect the way we live. It should be clear that any device with these characteristics will be a profoundly capable being, one much more capable than any existing robot or computer system. The technology required to create it is complex, difficult, and not available—yet. As will be seen in the following chapters, we have a few years to consider what such a development will mean to us, although probably not as many years as you might think. It is good that we have this time, however, because the development of an intelligent android raises an enormous number of vitally important issues that each of us, and society as a whole, must consider. The most obvious and commonly expressed thought is that intelligent androids will free people from many kinds of dangerous tasks, or release people from the drudgery of dull and boring jobs. Such an opinion is almost certainly true, but other implications are much more serious.

For example, what constitutes intelligence? How will we know that the android we build is truly intelligent? Will we be able to

*The distinction between the two terms is merely that the robot may or may not have a humanoid appearance, while the android certainly appears quite humanlike. In addition, an android may be constructed partially of biological materials, whereas a robot is strictly mechanical in nature. For now, the use of the word "robot" in this book implies a nonhumanoid, mechanical device such as a robot arm; "android" describes an intelligent device that appears more or less humanoid.

control its actions? If it is truly intelligent, it ought to be able to display some level of judgment and discrimination; what if its judgments don't match ours? What do we do then?

If an android performs an action that results in the death of a human, is it guilty of murder? Or is its owner guilty of murder? How should society deal with such a situation? If a person performs an action that destroys or seriously disables a android, is the person guilty of murder? Or merely guilty of vandalism? How should society regard that situation?

More than these, however, many, many other questions must be considered. As androids become more and more capable, they will certainly approach the intelligence level of a human being. Once that happens, whether it be twenty or two hundred years from now, still more questions arise. For example, should it be legal to own an android that has an intelligence level that equals (or exceeds) that of an average human being? Is that not slavery? If not, why not? Would an intelligent android have feelings? Could it experience emotions? Could it feel pain? Or happiness? Would such a being be alive? How could we tell, one way or the other?

What moral tenets should govern our dealings with androids? Will it be legal or moral to "turn off" an android? Under what circumstances? Is it moral to order an android to perform a task that will certainly result in its destruction? What do we do if it refuses the order?

How would the development of a functioning, intelligent android affect religion? Would success at this venture not imply that we are on a par with God? Would the androids worship us as their Creator? Why not? How will we deal with this? Will they even have a God? Do they have souls? Again, how could we tell one way or the other?

Where do we draw the line between considering an android as a machine that is property and considering it as an independent, intelligent being that should have some control over its own fate? Almost certainly, no one will argue that the first housekeeping android is anything but property, a mere machine. Yet as androids get smarter and smarter, their capabilities continuously advancing as our technology advances, when will we be forced to concede that they have rights just as we do? And how should society adjust to them, given the inevitable social, economic, and religious repercussions that will come if and when they are acknowledged as independent beings? Will androids fit into human society, or will they construct a society of their own? And how would a separate android society affect our human lives?

Questions such as these have no answers right now, only opinions. Until now, the answers have been mere academic arguments, with little relevance to daily life. But as our technology advances, and working androids come closer and closer to reality, these issues will become critical questions with which our society must deal explicitly and openly. The development of intelligent androids will affect us in many ways: legally, socially, economically, ethically, and morally, to name just a few. As we are forced to come to grips with the implications of our own creations, I believe we will learn much about who we truly are and what it really means to be human. The lesson may be hard-won, as with many that our species has learned, but at the least it should be illuminating.

In later chapters, I will attempt to sketch some of the arguments on both sides of these questions. Undoubtedly, proponents of all views will believe their argument has been short-changed or misinterpreted. To those people I can only apologize in advance. The purpose of this book, however, is not to be a polemic for one side or another; instead I will consider it successful if it causes even a few people to consider these issues carefully, and begin to decide for themselves how society should deal with the social, legal, and moral revolution that intelligent androids will cause.

Two key technologies will arise repeatedly, and it is a good idea to know something about them before we begin. These technologies are artificial intelligence (AI) and neural networks. A brief, introductory overview of each follows.

Artificial Intelligence (or "AI")

AI can be defined as the effort to reproduce the intelligent behavior of people through logical, step-by-step procedures. AI seeks ways to make digital computers more intelligent in their behavior by taking advantage of their deliberate, step-by-step execution process to mimic the step-by-step process of logical reasoning.

For instance, the problem of understanding language—parsing a sentence, for example—is one in which explicit rules can be used to generate a solution. All languages have a grammar that contains the rules on how to construct meaningful sentences. Sentences can be broken down into phrases and words, and their meanings constructed using those rules, along with a knowledge of the world-context of the sentences. Such natural language understanding techniques ("natural" language because it deals with ordinary human language rather than artificial languages like those used in computer programs) make up one large area of AI research. In effect, the AI notion is that other kinds of problems as well as language understanding also have a sort of underlying grammar which, if only it can be discovered, will lead to a step-by-step solution of the problem at hand.

Clearly, many of the rules used to solve everyday problems are not hard-and-fast rules, but instead are "rules of thumb" or heuristics—if we don't know exactly what to do, following a rule of thumb will probably get us closer to the answer than we were. Just as clearly, such rules of thumb don't always work. For example, suppose an android is standing at the top of a hill and is supposed to go to the bottom. There are no obvious paths down from where it stands, so it can go in any direction it chooses. In which direction should it move to get to the bottom most efficiently? A rule of thumb can help solve this problem: It should always move in the direction that is most steeply downhill from its current position. (This ad hoc solution assumes, of course, that there are no cliffs to walk over; while doing so would certainly get it to the bottom efficiently, it might also prove to be just a tad too efficient for the android's health!) Suppose we program the android to follow this heuristic. It is instructed that whenever it wants to get to the bottom of a hill it should always take the steepest (navigable) downhill path. Will it always be able to find its way down the hill to the bottom?

Unfortunately, the answer is no. Suppose that halfway down there is a culvert or gully that circles the hill. The android takes the most efficient route from the top of the hill to the bottom of the gully. However, once in the gully, it is unable to continue; in order to get all the way to the bottom it must first climb out of the gully and only then can it continue downward. The simple rule of thumb the android uses proves to be inadequate to solve this particular problem. Chances are, by the way, that a person stuck in a similar position would quickly find his way out of the gully and continue down the hill. This is because people are notoriously good at knowing when a rule has broken down and thus when it is necessary to change the rule or invent a new one. In other words, people are excellent problem-solvers. The difficult task is to provide androids with similar problem-solving skills.

Defining the field of AI can be tricky because so many different applications fall under its umbrella. One way to describe AI, however, is by the kinds of problems it attempts to solve. There are two separate categories of problems within AI: fundamental issues and final applications.

The list of AI applications includes problems such as general robotics issues, natural language understanding (comprehending the meaning behind colloquial English), speech understanding (correctly interpreting sounds as a sequence of words and sentences), vision, machine learning, planning, general problem solving, and a host of other similar problems. If you detect a similarity between this list of AI problems and the list of android characteristics stated earlier, it is no accident. AI research affects nearly every one of the requirements listed earlier.

From another perspective, however, AI's fundamental issues include any problem that can be widely applied to many of the specific AI applications mentioned above. For example, one of the basic problems nearly every AI application has to contend with is the difficulty of representing knowledge in an appropriate form. Suppose we need to train a robot arm to manipulate a set of blocks of various sizes, shapes, and colors. It would be nice to be able to give it a general direction such as "put the blue ball in the yellow box" and have the system generate a planned sequence of motions to carry this out. The robot might have to open the yellow box, remove some other object to make room for the blue ball, pick up the ball, put it in the yellow box, and close the box. The problem is that for the robot to be able to construct such a plan implies that it must also have some knowledge about the objects in its world—that boxes can be opened, for example; that they have a limited capacity; and that balls will roll away if not placed carefully.

The difficulty in representing this kind of knowledge in an intelligent program is that the system needs to store and retrieve such data in a highly flexible fashion, whereas in a computer, stored knowledge is generally accessible only if it is told exactly where to find it. Special matching rules and other techniques that might be used to find the needed information are often slow and cumbersome. People don't have to review every detail about blocks and balls and boxes in order to know how to manipulate them, and we don't want an artificially intelligent system to have to do that either. Also buried in this issue is how to represent the problem itself and the answer that the system is to generate. How exactly is the program to know what it is supposed to do? How is the goal of the system to be represented in a manner compatible with the knowledge that the system has? And, finally, how can it match the perceived world around it to its knowledge and goals? These questions are quite difficult to answer, and no general solution has yet been formulated that succeeds in all problem environments.

Other fundamental issues exist in AI as well. One is constraint exploitation. This means that when a problem has known limits on its solution, the system should take advantage of these limits when it is searching for an answer. For example, if the problem is to build the highest possible tower out of a number of differently shaped blocks, a person will automatically ignore any solution that calls for putting a round ball on the bottom of the tower—people know from experience that little or nothing can be stacked successfully on top of a round object. The implicit constraint that the tower has to be at least reasonably stable thus causes a person to ignore many potential, though impractical, solutions. An intelligent android should also not bother to consider putting a ball at the bottom of its tower. Another example might be the problem of finding an efficient path through crowded city streets. An experienced driver would know to avoid routes with substantial amounts of construction; an autonomously directed vehicle should do the same. The kind of problem often resolves into a problem of determining the most efficient means of searching through all possible solutions to find the best—or at least a very good—one. In fact, some researchers consider the development of intelligent search strategies to be the most fundamental AI problem of all.

Another fundamental issue in AI is that of using logic to solve a problem. Very often, AI systems implement a form of logic called predicate logic, which formally states facts about objects as predicates in a sentence. "All men are mortal" is an example, with "are mortal" being the predicate that describes "all men." Some AI practitioners have even developed a special programming language called Prolog (from "*pro*gramming *log*ic") that represents complete computer programs as a series of predicates and that follows rules of logic to resolve these predicates during program execution.

Still another major issue in AI is that of problem solving. How is it that people can solve a problem they have never seen before? What mental contortions do they go through? One of the key research areas in AI is to develop techniques that mimic this human ability for at least certain kinds of problems. Several successful methods have been found. One method is to generate a set of possible answers (or partial answers) and test each to see which works best. This usually results in an extended list of possible solutions, so implicit in this method is an ability to search through the list in some reasonable fashion and prune it down intelligently in order to limit the solutions that must be tested and so solve the problem efficiently. In many problems, some of the potential solutions, such as any tower-building solution that places a ball at the bottom of the tower, can be immediately rejected without testing them; others can be rejected because of other obvious drawbacks or through the application of heuristics. This reduces the total number of possible answers that actually have to be tested to a more reasonable number.

Yet another extremely successful AI technique is to develop an expert system that explicitly encodes the rules governing the environment the system must deal with. Rules are typically encoded as a collection of "if-then" statements: *if* the sky is blue, *then* leave your unbrella at home today. These rules, along with known facts about the current state of the environment, constitute a rule-base and a knowledge-base. The heart of such an expert system is an inference engine that compares the if-clauses of the rules in the rule-base to the known (or believed) state of the world to generate a collection of rules that match this state. One (and usually only one) of these rules is selected by any of a variety of techniques, and the resulting then-clause of that rule is executed—that rule is "fired." Presumably, the effect of this rule firing is to change the state of the world slightly so that a new collection of matching rules can be generated. For example, the result of executing the rule mentioned above might be that you leave your house with no umbrella. The new "state of the world" is that the umbrella is still in the umbrella stand at home and you are away from the house with no protection from the rain.

This summary of AI is necessarily brief, but the fundamental characteristic of the field can be expressed as follows: AI is the attempt to make computers behave intelligently, particularly in real-world situations.

Neural Networks

The second major technology that will appear repeatedly in this book is that of neural networks. Neural networks are information-processing systems physically structured in a way that mimics our current understanding of the brain. The architectures used in neural networks are now little more than crude approximations to the brain's physical structure. Even these simple models, however, can accomplish some astonishing feats.

A neural network is not a computer in the ordinary sense, for it does not compute anything, nor does it run a program or follow a sequence of instructions. Instead, a neural network reacts to an input stimulus by generating a pattern of activity in the network. This pattern of activity is transmitted through the layers of the network from the input side to the output side in a process called "spreading activation." The activity pattern that appears on the output side of the network is its "answer" to the input stimulus.

A digital computer, in contrast, typically is made up of several distinct units. This usually includes one or more physical boxes that contain a keyboard (or other device such as a mouse) for input, a video monitor for output, and a processing unit. (Often, of course, several other devices like disk drives or printers are attached to the computer, but they are not important for this short overview.) In a digital computer, the processing unit is divided into at least two pieces, a memory and a central processing unit (CPU).

The memory consists of a collection of semiconductor chips (or other devices) that stores the data the computer will work on, as well as the specific, step-by-step instructions that it will perform. Information is stored in the memory in a series of labeled pigeonholes, called addresses. Each pigeonhole can contain a single instruction or a single piece of data. To recall any given fact or instruction, the computer must be told exactly which pigeonhole contains it; then it can retrieve that item and either operate on it (if it is data) or execute it (if it is an instruction).

The CPU is a set of one or more chips that actually performs the instructions stored in memory, and it does so in a very special fashion. All digital computers execute their list of instructions—their *program* —in a three-step process: (1) they retrieve the next instruction to perform, as well as any data needed for the instruction, by fetching it from a specific memory location (usually the location specified in a particular memory position called the program counter); (2) they execute the instruction; and (3) they store the results of the instruction, if any, back into memory in the assigned memory location. Every digital computer today follows this basic fetch-execute-store sequence. It is repeated thousands or millions of times a second as the computer steps through its list of instructions to complete a program. The entire process is repeatable, predictable, and well understood. As the saying goes, there are no bad computers, only bad programs.

Neural networks, in contrast to the digital computer, are designed to imitate the structure of the brain, specifically that of the cerebral cortex. They generally fall back on a very simplistic model of the brain, however, because such simple designs are easy to implement and test. In the most commonly used model, the brain is considered a collection of neural cells, or neurons, that have many connections to other neurons. An abstract neuron in this model consists of a cell body that has a number of filaments protruding from the top (input) and a single long filament extending from the bottom (output), as shown in Figure 1.2. The top filaments, called dendrites, receive signals from other neurons or sensory organs. The bottom filament, called the axon, transmits the neuron's signal to other neurons' dendrites. The axon can branch hundreds or thousands of times, creating axon collaterals, with each branch terminating at a dendrite of some other neuron or, possibly, terminating at a muscle or other "output" device in the body. A signal transmitted by the body of the cell down the axon will eventually work its way along every axon collateral to end up at many destinations. Note, by the way, that every axon

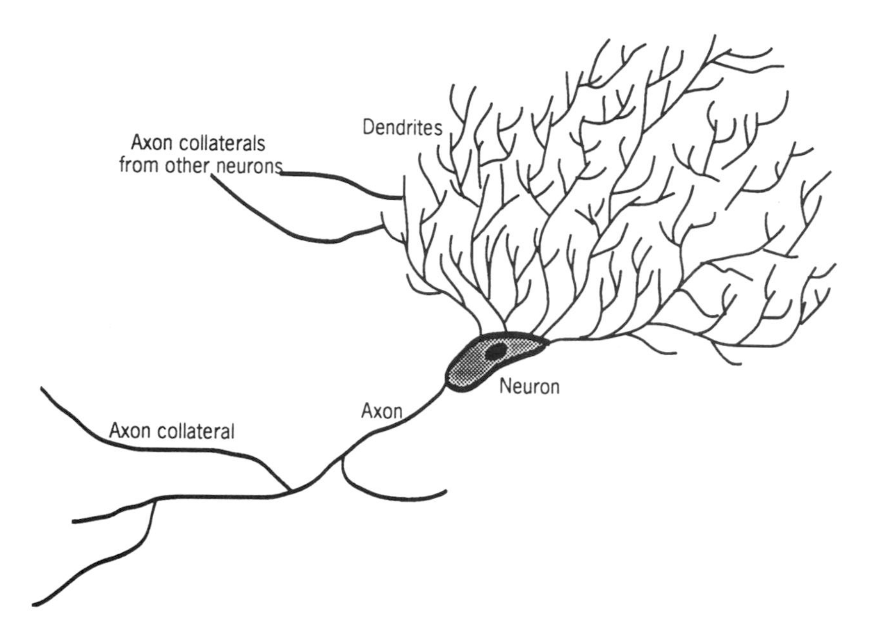

Figure 1.2 A typical neuron. Each neuron has many dendrites to receive inputs from other neurons. In contrast, it generates a single output that it transmits over its axon and along the axon collaterals (branches).

collateral transmits the same signal value to its eventual terminator; the signal is not split or modified in its journey out of the neuron. This is a vastly oversimplified version of true biological neurons, of course, but it suffices for now.

A neural network's structure is based on this simple neural model, and is quite different from the neatly compartmentalized design of a digital computer. A neural network consists of a large number of tiny local processing nodes, or neurodes, that are individually quite simple, but have many connections to other neurodes. Like the crude neuron model above, each neurode has many incoming connections, but only a single outgoing signal. This outgoing signal may branch many times, so that the signal itself goes to a lot of other neurodes, but in all cases, it is the same signal going to each destination, rather than a set of different signals, or a portion of the signal going to each. And like the brain's neurons, the neurodes of a neural network have from dozens to thousands of connections to other neurodes, to sensory devices, and to the outside world.

A neurode does perform certain computations, but these are very basic ones indeed. And in so doing, it follows none of the computational steps used by a digital computer in processing information.

Typically, each neurode computes a single mathematical function, called a transfer function. In computing this function, the neurode first assesses the total level of incoming stimulation from all of its input connections, weighted by the individual strengths of each connection. For example, a moderate sized signal arriving along a very strong connection contributes a larger share to the total input than the same sized signal would arriving along a very weak connection. All these individual contributions are added together to make the net input signal.

This incoming stimulation is then translated through some nonlinear relationship (in other words, a relationship not expressed by a straight line) to an internal activity level. Most commonly, the nonlinear relationship is implemented by a sigmoid function, or one that has an *S*-shaped curve.*

The neurode then uses the specific internal activity level derived from the sigmoid function to generate a corresponding output signal, which is then transmitted over the neurode's outgoing connections. This is often a simple threshold that merely serves to ensure that the neurode does not respond to tiny incoming stimuli. If the neurode's activity is less than the threshold value, nothing is output; if the activity is greater than the threshold, the activity level itself is the value output. This guarantees that each neurode's output will be of significant size, and that the network will not be flooded with tiny, meaningless signals. (As a result, by the way, the output signal has a maximum value because the sigmoid limits the neurode's activity level.) This "input stimulus-internal activity-output signal" process is the neural network equivalent of a digital computer's fetch-execute-store.

What is still more interesting is that typically all neurodes in a neural network (or, in very complex networks, all neurodes in a particular subsection of the network) compute the same transfer function. Neurodes do not have separate tasks to perform or miniature programs to run, as would be true in a parallel digital computer. Furthermore, the transfer function they compute has absolutely nothing to do with the nature of the task that the network itself is performing. In fact, you could take a neural network and train it to, say, generate speech, and then, without modifying the network or the

*A sigmoid is a function that has a low, shallow slope initially, followed by a steeper rise that gives way to another low, shallow slope. Its value always increases as its input increases, yet, at the same time, it is bounded by maximum and minimum values. Neurodes usually use any one of a number of sigmoid functions to determine their activity level for a given total input stimulus.

neurodes in any way, retrain it to do image processing. Nothing about the neurodes or their transfer function needs to be changed for them to accomplish such distinctly different tasks.*

Information processing in a neural network takes on a totally different aspect from that in a digital computer. In a computer the steps by which information is assimilated and processed are completely identifiable. The data itself, along with the program that will handle it, are easily retrieved from memory, and they can each be assessed for correctness. In a neural network, however, information is processed quite differently. An incoming stimulus pattern is presented to the network. Some of the neurodes in the network receive this stimulus pattern directly. As a result, they become stimulated and active to a greater or lesser degree. Because each neurode receiving the pattern has its own individual set of connections of varying weights, the activation level of each of these neurodes will vary as well. Each responds with the appropriate output signal that branches out to all the neurodes connected to it. This pattern of activity may differ considerably from the original input pattern; it depends primarily on the relative strengths of the connections between the outside world and the network.

The neurodes that receive the various signals from the first level of neurodes will themselves be activated to a degree that depends on the strengths of their incoming connections. As a result, a new pattern of activity is generated within these neurodes, and a new collection of output signals transmitted along the network's connections. Eventually this pattern of spreading activation reaches those neurodes that transmit their output to the outside world; the activity pattern in these neurodes is the network's response to the input stimulus.

In effect, an input pattern presented to the "input side" of the network comes out the "output side" as a pattern of activation. The processing that occurs is a result of the physical structure of the network, the transfer functions of the neurodes, and the pattern of connections and their strengths. There is no way to retrieve the "program" used to generate the network's response, because no set of specific instructions was actually performed other than the transfer functions within the neurode, which has nothing to do with the network's overall task. A neural network's method of processing data is thus more of a knee-jerk response to a stimulus than a reasoned answer. Like the brain, the network's methods and techniques must

*At worst, you might have to modify the physical size of the network, particularly the input and output layer sizes, to make the network able to cope with a new problem. The neurodes themselves and the general connection scheme, however, would not have to be changed in the slightest.

be inferred from its actions rather than analyzed in a step-by-step fashion.

It turns out that the real work of a neural network is not done within the processing elements, as might be expected, but instead is performed by the connections between the neurodes. These connections have a characteristic weight or strength associated with them. This strength is a bit like a handclasp: The junction can be as flimsy as the lightest touch of a finger, or as sturdy as a firm grip. A weak connection means that signals have difficulty in transmitting between two neurodes; a strong connection implies that the signals transmit easily. Furthermore, these junctions can be either excitatory or inhibitory. In an excitatory connection an incoming signal tends to increase the receiver's activity, and thus increase its output; a signal arriving through an inhibitory connection tends to lower the receiver's activity, and thus lower its output. Usually some of each neurode's incoming connections are inhibitory and some are excitatory.

The most interesting characteristic of the connections in a neural network, however, is that the strength of these connections can change. In fact, a neural network is trained (not programmed) by systematically altering the weights of its internal connections until the network behaves as we wish. (The transfer functions of the neurodes usually are not significantly modified, if at all.) This process is called training rather than programming because typically it is performed by presenting the network with a series of example inputs, watching its output for each, and adjusting the weights on the connections to improve the network's performance. When the performance is good enough by whatever standard we choose, the network is considered fully trained. The process is perfectly analogous to that of a tutor training a student in, say, recitation or music.

Whatever the network knows, the information that it uses to perform a task, is stored not in the pigeonholes used by a computer, but in the pattern of these changing weights. Training, and thus learning, in a neural network consists of changing the weights on the connections between neurodes, not changing the neurodes. The real "intelligence" of a neural network is in the connections and pathways, not the neurodes themselves.

In the late 1950s and early 1960s neural networks were a very popular research system. At that time they were commonly called perceptrons, and some scientists had extremely high hopes for them. In particular, one misconception was that if we could make a large enough collection of perceptrons, say about a hundred billion of them, and if we randomly connected them, placing a few tens of thousands of connections on each one, and if we then just "turned it on," the result would be a brain just like the one in a human being. Of course this is not the case. The human brain is a highly structured and organized device, far more complex than this sort of random interconnection would imply. Also, the subtleties and complexities of both neurons and their connection structure are vastly greater than the simple perceptron could imitate. The point here, however, is that even thirty years ago some scientists thought that neural network technology might someday enable them to create an artificial brain.

Neural networks clearly differ very strongly from the traditional approach taken by most AI researchers. AI is an attempt to make computers behave intelligently; neural networks give up even the physical structure of the computer and instead try to develop a new and highly parallel architecture that behaves intelligently without reasoning through a problem to a solution. In some sense, it can be argued that workers in AI are rationalists, believing that intelligence derives from logic and reason, while neural network researchers are emotionalists, believing that intelligent behavior often derives from instinctive or unreasoning responses to situations. While this analogy implies some extremes that are unfair, its basic premise is sound: AI is based on logic and reasoning while neural networks are based on unreasoning reactions to stimulation.

Why are neural networks important in building an android? The answer to this question lies in the kinds of problems they are good at solving. In general, neural networks can solve problems that fall into the categories of pattern matching, categorization, functional mapping, and process control. Let's see what each of these means.

Pattern matching is the capability of correlating an overall pattern of input signals of any type with an output signal of some arbitrary type. For example, when a person recognizes a friend's face, the skill used is that of pattern matching. This is so because the friend has almost certainly never been seen in exactly that fashion before, with hair, expression, and pose exactly the same. Yet people are easily able to match this new image of the person with the friend's name, and do so immediately and without conscious effort. Also in the realm of pattern matching is pattern completion; this is the ability to fill in missing or incomplete patterns so that the entire pattern is perceived rather than a partial pattern. This occurs, for instance, when we recognize someone from only a partial view—such as when we recognize Mickey Mouse by the shape of his ears. Again, this skill is so unconscious that people often are not even aware that anything was missing in the pattern; neural networks may have this capability as well.

Categorization means the ability to group objects and concepts by their relative similarities or differences. This is the basis of similes and metaphors, for example. "My love is like a red, red rose" is clearly a categorization of "my love" with a flower. Yet on a much less esoteric basis we do this all the time. If asked to name parts of the body, the average person has no difficulty in listing "foot, arm, head"; if asked to name nautical terms, a similar list of "aft, head, galley" can be recited. And if asked to give words associated with a glass of beer, "head, alcohol, hops" come to mind. The word "head" can successfully be categorized as a part of the body, a part of a boat, or an attribute of a glass of beer, depending on context and need. Similarly, we categorize people as "good guys" and "bad guys," and cars as "fuel efficient" or "gas hogs." Categorization is so innate to the way people think, and so useful as a sort of mental shorthand, that we hardly know we are doing it. Like people, certain neural networks have shown strong abilities to lump patterns into categories, and sometimes even to change categories with changing context.

Functional mapping is another general neural network application. A mathematical function is a relationship between two sets of numbers such that for each number in the first set there is exactly one number in the second set. The first set of numbers is called the domain of the function; if we draw a graph of the function this set of numbers is on the horizontal axis. The second set of numbers is the function's value for each corresponding number in the first set; this lies on the vertical axis of the graph. The mapping merely means that a relationship exists between, say, 3 in the domain set and 9 in the range set. A similar mapping might exist between 2 in the domain and 4 in the range, and between 5 in the domain and 25 in the range. Sometimes the relationship can be summarized with an equation; in the example above we might suspect that the function is equal to the square of the domain value (i.e., $2^2 = 4$; $3^2 = 9$; $5^2 = 25$). Not all functions are simple or even expressible in terms of an equation, however. There are some relationships in which no known equation exists that correctly describes that mapping from the domain to the range, or in which the equation is so difficult and so complex that computing it is extremely hard or impractical. For these kinds of functions the neural network ability to model the functional mapping is useful. A neural network learns the mapping based on a collection of examples of the relationship, rather than requiring the formal equation as a computer would.

Process control is the final general neural network application. This is an application where the problem is to view the state of something and maintain that state in an appropriate fashion at all times. The neural network acts as a sort of watchdog in this case, adjusting the process appropriately to keep it working well. One of the best examples of process control is cooking a pot of spaghetti sauce. A good cook will constantly check on the sauce, adjusting the heat of the burner, stirring the pot, adding an herb here or a spice there, until the taste is just right. The cook that does this is performing a process control application by adjusting the cooking process appropriately to generate the desired output. Process control applications cover everything from cooking a pot of spaghetti to controlling a nuclear power plant. Even the general field of robotics can be considered as a specialized process control application.

These, then, are the tools we have to work with in the attempt to construct an intelligent android; we already know the requirements we need to meet. Are these tools adequate to achieve that goal, and if not, how long will it take to refine them? The next chapters consider this question in detail, beginning with one of the most critical systems in the android: the sense of vision.

≧ 2 ≦

There Is None So Blind

I have never seen a greater monster or miracle than myself.

Montaigne

There is a story about a starving artist who struggled for many years to make a living with his paintings. He created ugly, depressing works in shades of black, dark gray, and dark brown that received some critical praise, but were far too gloomy for most people to actually buy and hang on their walls. He barely scratched out a living.

One day the artist decided that the world was not well lost for the sake of his art, and he approached his agent for suggestions on how he might make more money. The agent suggested that he lighten up and put a bit more color in his paintings, so that people would be able to look at them for more than a few minutes without becoming depressed. The artist listened carefully, and mused on the advice all the way back to his garret.

A few days later, he returned to his agent's office with his newest painting. The agent uncovered the painting and stared at it in shock. The picture was of a childishly drawn daisy in a vividly red vase, seen against a bright blue sky. It looked much like the kind of art sold in hotel rooms for $15.00 per "original artist painting." "It's nothing but schlock!" the agent complained.

The artist smiled and nodded. "Yes, it's schlock," he agreed, "but it's *colorful* schlock."

While the artist may have taken his agent's advice to extremes, there is no doubt that people react strongly to what they see. We are very visual creatures, depending on vision to an extraordinary degree. Whereas some animals rely on acute senses of smell or hearing, a human's abilities in these areas are no more than mediocre. Human beings do, on the other hand, have exquisitely sensitive, full-color, stereoscopic vision.

Not only is vision necessary to us as a means of dealing with the world, it also strongly influences our attitudes and behavior. For example, one of the hottest consulting businesses in recent years is that of "color consultant." Such people advise their clients on appropriate colors to wear, decorate with, or use in different ways to make varying impacts on their lives. While it is not completely clear how useful such expertise is in general, certainly what we see affects our moods and attitudes very strongly.

Vision is more than just having an image projected on the back of a projection screen in the eye. In fact, we do not yet fully understand exactly how our visual system works, and much of what we do know has only been discovered in the past ten years or so. Still, scientists are beginning to piece together how we go about understanding what we see. And the answers they are coming up with are quite different from what they thought they would be originally.

Until very recently, the common notion of how we see had more to do with mysticism than with fact. As nearly everyone knows, the eye resembles a simple camera, with a lens at the front, and a "projection screen" at the back inner surface. This projection screen is the retina. The basic (and wrong) idea of vision that prevailed for many years was that the lens collects and focuses light rays from an object, and projects it onto the retina. This complete image is then transmitted via the optic nerve up to the visual cortex area of the brain, where it is looked at and interpreted by (choose one) (a) the soul; (b) the mind; (c) a little man, or *homunculus;* or (d) something else. This actually sounds like a perfectly reasonable explanation of vision—except for the fact that it is wrong.

One of the problems with this theory is that the lens in the eye is a simple, convex lens, and it has the result of inverting the images it focuses onto the retina. Figure 2.1 illustrates this phenomenon. The reason inversion of the figure causes difficulties with this theory is that no one could really explain how people eventually end up perceiving objects as right side up, when the images always arrived at the brain upside down. The most common explanation offered was that we really do see the world upside down, but because an infant learns that when he lifts a hand, it really moves upward instead of the inverted downward motion that his eye sees, he merely learns to *interpret* the upside down images as rightside up. Once past this learning phase in early infancy, re-inverting the already inverted images becomes so ingrained that people can't really tell that they are actually viewing life upside down.

The problems with this are both that there is no obvious mechanism in the visual system that would invert an image like that, and

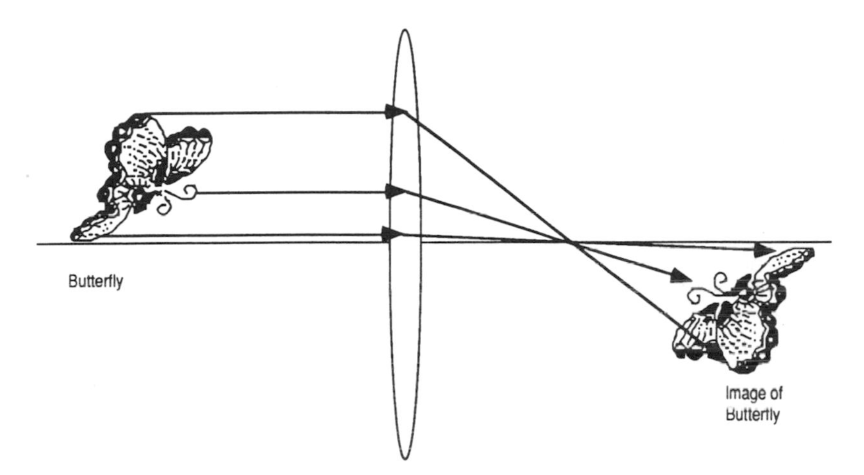

Figure 2.1 A simple lens such as that found in the eye inverts the image of an object.

there is also no central homunculus that can do the viewing. Vision just isn't quite that simple.

So how does vision work, if it isn't like a kind of movie projector system? First of all, the basic idea of how the eye works outlined above isn't too bad. The lens really does collect light from an object and focus it on the back interior wall of the eyeball, where the retina is located. But from here, the processing is very different indeed.

The retina is composed of a large number of light-sensitive nerve cells, called rods and cones from their general shape. Rods are the most sensitive of the two kinds of receptors; they can perceive even a few photons striking their surface. Because they are long and skinny, they can also be packed together very tightly, allowing high-resolution details to be observed. Their physical compactness allows about 100 million of them to be crammed into a retina about the size of a postage stamp. Rods are sensitive only to the presence or absence of light; they "see" in black and white only.

Cones, on the other hand, are shorter and pudgier than rods, so they can't be packed nearly so tightly together; only about 6 or 7 million of them will fit in the retina. To make up for their resulting lack of resolution, cones can perceive differences in color. There appear to be three different kinds of cones in the retina, although they do not seem to be organized solely as red-perceiving, blue-perceiving, and green-perceiving (red, blue, and green being the primary colors of light). Cones are also not as sensitive as rods; when the light is very dim, color vision is lost and the twilight world becomes black and white.

The retina has rods and cones bunched together in a tightly packed sheet that covers the interior back wall of the eyeball. Cones tend to be concentrated near the middle of the field of vision, while rods are present throughout the retina. The further from the center of vision, the fewer cones are present, with the result that peripheral vision has very little color sensitivity. On the other hand, with the high concentration of rods in the peripheries of the retina, the eye is extremely sensitive to changes in that section of the visual field. In other words, we are exquisitely aware of motions that occur at the edges of our vision—a useful characteristic indeed for a creature trying to avoid attacks from predators. It happens, by the way, that because of the higher concentration of rods slightly away from the center of the field of view, we can see better at night when we don't quite look directly at an object. Astronomers and stargazers are familiar with this phenomenon; they generally look just beside a particular star or planet in the sky to see it most clearly. (Telescopes and binoculars, not having either rods or cones in their systems, should obviously be pointed directly at the desired object of study.)

The rods and cones in the retina are just like any other neural cell in that they become excited when stimulated. The number of photons (light particles) that fall on the receptor of each cell in the retina determine how excited that cell will become. But it is not true that the retina acts as a kind of projection screen. Instead, the retina is the first step in a complex image-processing system. Just as a computer considers an image to be made up of tiny picture elements (pixels, for short), the rods and cones in the retina act as a natural means of breaking down an image into variable-sized pixels that correspond to the locations of each cell. Where the cells are denser, the pixels are smaller and more detailed; where they are less dense, the image loses resolution (but may gain in color information, for example).

The neural receptors in the eye, like any others, exhibit a property called saturation. This means that if they are constantly stimulated with the same input signal, sooner or later they will stop responding to that stimulus. For example, if you have ever stepped into a pool or the ocean and found the water too cold, only to realize that it is comfortably warm after you have had a chance to "get used to" the temperature, you have experienced saturation. The neurons that detect cold saturate after a few moments in the water and can no longer transmit the sensations that cause your brain to make your body shiver; the result is that the goosebumps go away and you feel perfectly comfortable.

You can also experience saturation by going into a room with large blank walls painted some plain color. Stand very close to the wall so that everything in your field of view is a fixed color. After a few moments you will find that the wall appears to have changed color; you can no longer see its original hue. A green wall may appear to become pink or purple, for example. This is called a negative after-image.

When doing this experiment, you must stand very close to the wall because otherwise the natural motions of the eye, called saccadic motions, allow the retina's neurons to get a rest from the color. These saccadic motions occur constantly, without our being aware of them. Essentially, the eye constantly "dithers" so that no single spot on the retina is focused for very long on exactly the same position in the image. This helps prevent saturation of the neurons in the retina, even when you stare at a single spot—that fixed stare is not nearly as "fixed" as we believe. Generally speaking, saccadic motions are so slight as to be unnoticeable either to us or to an observer. Occasionally, however, an individual's normal saccadic motions are large enough that their eyes seem to constantly quiver to the onlooker—one well-known actor took advantage of this to make a devastatingly intense and sexy stare in a film version of *Dracula*.

If there are three kinds of cones, yet they don't appear to be only sensitive to the standard red-blue-green colors, what *do* they do? Scientists don't have an absolute answer to this as yet, but one current theory seems to have a lot of evidence to support it. This is the opponent processing theory. Understanding it may help illustrate the fact that the projection screen notion of vision is definitely wrong.

The cones and rods of the retina do not directly transmit their signals to the brain. Instead several layers of neural cells immediately behind the retina perform some preprocessing of the image long before it arrives at the brain. In particular, one group of cells seems to deal with color perception. The opponent processing theory says that there are six different kinds of cells in this particular group. Two of these types are sensitive to either red or green light. Of these, one type is excited when red is present and inhibited when green is present; the other is excited by green and inhibited by red. They are *opponents* because their reactions are exactly opposite. Two more kinds of these preprocessing cells are blue and yellow opponents, and the final two are black and white opponents. In each case, one kind of cell is excited by one color, while its opposing cell is inhibited by the same color. The idea is that color perception arises from the complex pattern of excitement and inhibition that arises from these processing cells. Notice that before the image has even arrived at the brain, the visual system has begun to break it down into its components and try to figure out what the pieces mean.

Animal experiments have been done that illustrate another aspect of the visual system. David Hubel and T. N. Wiesel won a Nobel Prize for their studies in the late 1970s in which they clearly demonstrated that certain neurons in the visual system look for specific visual features. These are grouped into simple cells, complex cells, and hypercomplex cells. Simple cells become very excited when they are presented with images of lines and bars in various orientations; one particular simple cell might respond strongly to a vertical line, for example, but not to a line in any other orientation. Complex cells do not so much respond to a particular pattern as to the motion of a particular pattern across the visual field. For example, one complex cell might become very excited if a horizontal bar moves from the bottom to the top of the visual field, but that same cell would have little or no reaction to a slanted line moving the same way. Finally, hypercomplex cells respond to more complex patterns such as corners, angles, or even complex general features such as the length or width of some object in the field.

These feature detectors help the visual system break down the image and process it so that it is understandable. They also control how much and what aspects of the image actually reach the brain for further processing. If there is no simple, complex, or hypercomplex cell that can respond to some aspect of an image, it is much less likely that we will be able interpret it easily.

The point of all this discussion is that the brain actually breaks down an image into its component picture elements, just as an image-processing computer does. This may be one of the few ways the brain really resembles a traditional digital computer, by the way, so we should appreciate it properly. Studying the way the human visual system works turns out to be a pretty good design specification for the visual system of an android. But if the vision system is like an image processing system, how does *that* work?

The real difficulty in designing a vision system is that all you have as input are a collection of light and dark pixels. There is no information to tell you which group of pixels corresponds to a single object or what the scale and orientation of any object is. Furthermore, the objects we look at are not cartoons, with nice, neat black outlines drawn around them—at least not usually. They have fuzzy edges, they overlap one another, and often their edges are distinguished only by differing textures or colors. And even worse, we can't be sure that the pixels perceived are actually correct. Noise—the fluctuations in signal due to sensor error or other causes—is always present in any system, but particularly so in the biological impreciseness of the nervous system; this causes subtle and not-so-subtle errors in the relative strengths of pixels in the image. Because of all these factors, figuring out the meaning behind the pixels is challenging indeed.

We have already seen one way this can be countered. The color perception preprocessing that occurs just behind the retina is only one step in the early processing of an image. Early processing refers to computational steps that have the goal of analyzing the entire image. In this phase, the system determines what features and edges are present in the image, and generally treats the entire image the same way, processing it in order to determine the basic features within it. Early processing is performed best when it takes place locally and in parallel. In the eye this is performed, for example, by the opponent cells used in color perception. Sets of these cells are duplicated extensively, so that each set only has to concern itself with a small section of the image. By having the sets work in parallel, the total time needed to process the entire image is dramatically reduced.

Once the fundamental features within the image are determined, processing continues with a step called late processing. While early processing is a decompositional phase where the image is broken into its component features, late processing is a compositional phase during which the features are recombined into recognizable objects. Because late processing does not have to deal with individual pixels but only the features represented by those pixels, late processing is often performed sequentially in an image-processing computer system. While there may be half a million or more pixels in a high resolution display image, there will likely be only a few dozen or so features within that image that the late processing system has to interpret.

This is important, because the visual system is one that absolutely has to operate at real-time speeds since our eyes are constantly bombarded with new images all the time. But what is "real-time speed" in vision? The average person seems to have a "frame-rate" of about one-fifteenth of a second. That is, changes that occur faster than that are generally too fast for the eye to perceive. Movie projectors typically run at about 15 to 30 frames each second, allowing the action on the film to appear continuous to us. A television or monitor screen that refreshes the display every one-thirtieth of a second gives relatively flicker-free and stable images to most observers.

There is an interesting story associated with this frame rate. When the Apollo 11 astronauts were making the first landing on the moon, the onboard computers at that time were far less sophisticated than those available today. In particular, one computer tended to reset itself whenever it received too much data too quickly; every time it reset a light would turn on to inform the pilot what had happened. The engineers who designed the system assured the astronauts during training that as long as the system did not reset more often than every one-fifteenth of a second, the computer would still be able to perform necessary functions. However, if it reset more often than that, there could be trouble. When asked exactly how the the astronauts could possibly know that this threshold was exceeded, the engineers explained that as long as they could see the light flicker on and off, the system was all right, but if the light turned red and appeared to glow steadily, the computer was resetting too quickly. As it turned out, the computer did reset a great deal during that historic landing, a fact that worried Houston's Mission Control a lot. However, Neil Armstrong could see the red light flickering, so he knew that the computer was still able to do its job acceptably well.

In any event, ideally we want a vision system to be able to process an image about every tenth or fifteenth of a second. Typically, an artificial vision system uses parallel computers, neural networks, or array processors to carry out the early processing steps, and then may use AI or other techniques to handle the late processing stages.

Suppose an image undergoes early processing to determine the important features present in an image. Generally this means that the image is broken down into "edges," "bars," and "blobs." An edge is a boundary between two colors or textures; in black-and-white images it separates dark areas from light areas, or areas of one texture from another. A bar in image processing consists of two edges that are more or less parallel so that the region between them is separated from the lighter or darker region on each side. A blob is an area or region that is separated from the surrounding areas by edges; as the name implies such a region can be of any regular or irregular shape. These characteristics are generated by a complex system that makes a number of reasonable assumptions about what is being observed. For example, the visual system usually assumes that changes in texture correspond to changes in orientation. For example, Figure 2.2 illustrates a texture change that appears to be a surface slanted away from the page. Other assumptions also go into the early processing stages, many of which are the source of several optical illusions. Our vision system is not perfect, and the eye can be fooled, but the assumptions that go into this level of processing are reasonable ones based on those objects we are most likely to see.

The early processing system also handles image characteristics such as reconciling the different image views from each eye (binocular vision), dealing with the varying levels of illumination in the image, and processing the motion of objects across the visual field. Another important task is to determine the connectivity of points or edges in the image to see if there are any "virtual" lines. Figure 2.3 shows an example of an image that is interpreted as a series of lines, even though they are not actually connected; these are virtual lines. After all this is done through local, parallel processing, the system is ready to try to put the image back together and interpret what it has seen.

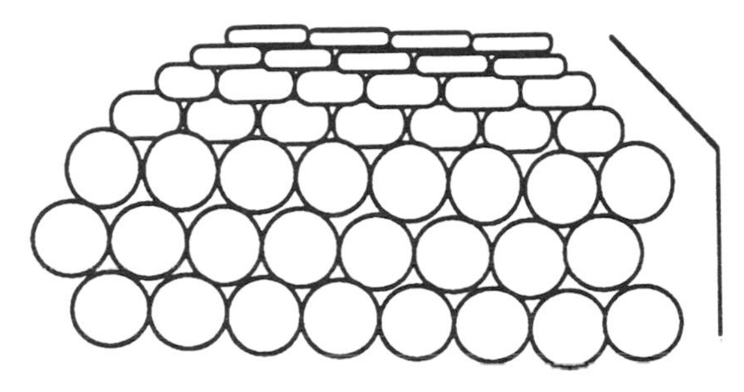

Figure 2.2 The changing texture makes the image seem to bend into the page as the line at the side indicates.

During the late processing stage, an internal representation of what is in the image is constructed so that it can be interpreted and stored in memory. The problem here is one of representation: What are the key elements in a scene and how do we structure those elements in memory? There is a serious problem here that must be addressed. Consider a small picture of a leaf. You might notice the veins on the leaf, specific shadings and textures of the leaf's surface, and so on. If you compare it to another image of a similar leaf, you are apt to note a number of subtle variations between the two. You probably say that the two images were distinctly different. Now suppose you take those exact same images of leaves and include them as part of a picture of a tree. Do you perceive the two leaves as being inherently different, or, more likely, are you inclined to state that all the leaves on the tree are pretty much the same? The small differences that were important when viewed at the level of the individual leaf are no longer so when viewed at the level of the tree as a whole. Thus, image representations must include the important features of im-

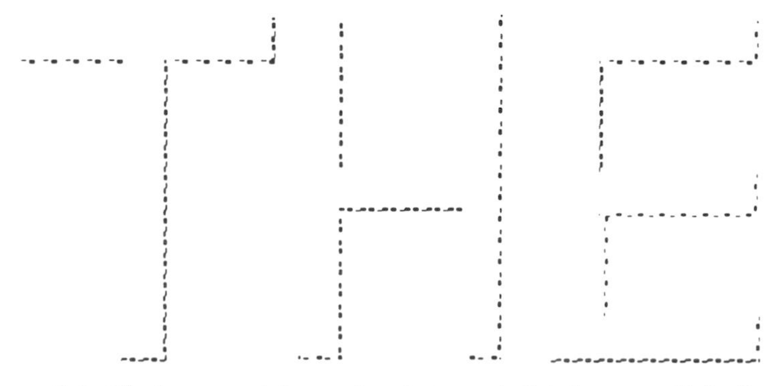

Figure 2.3 The human vision system interprets this image as if the lines are continuous and connected. It also ''fills in'' the missing pieces so that it is nearly impossible not to see the message written here.

ages, where what is important may change from moment to moment.

Other problems arise at this level of processing as well. Consider a child's block. The angle at which you view that block may change from directly ahead (so that its outside edges appear to be a simple square), to any elevation and orientation (so that its outside edges constitute a complex geometrical figure). No matter what the orientation of the block, you have to be able to recognize its inherent cubical shape. Similarly, no matter where the block is within the field of view, you have to recognize its shape and know that it is a block. This orientation and rotation invarient representation is essential in dealing with an unconstrained environment.

Or consider still another problem. When was the last time you saw a house? Unless it was newly constructed and had no landscaping around it, almost certainly you really did not see the house's complete outline. Intervening shrubs, trees, and other plant materials almost certainly concealed a significant portion of the surfaces and edges that compose the house itself. Yet no one has difficulty realizing that the wall that appears to stop just above the camellia bush actually extends on down to the ground. Or that the piece of wall to the left of the tree is connected to the piece of wall to the right of the tree. The representation of the image of the house has to take into account such details. This problem is called occlusion, and the house is said to be occluded (partially concealed by) the intervening plants.

Most efforts at image interpretation concentrate on storing objects by their collected features. For example, a tree might be defined as an object with a long, more or less vertical bar (the trunk), the top of which is wholly or partially occluded by small irregularly shaped blobs that are leaves; it might sometimes have smaller bars (branches) visible that extend from the main bar at various angles. To finish such a description, we might add that the normal height range for a tree is from, say, 6 to 50 feet tall.

The reason for this sort of representation is twofold. First, it requires far less storage space than a detailed pixel-by-pixel image of a tree. By breaking the tree down into general features, we use less storage than would be required if we were to store the detailed image. Second, having a general feature representation of the tree means that it is easier to recognize a new object as a tree. If each new example had to be a close match for the pixel image of our prototypical tree, the chances are that every single tree would be called by a separate name—even if we restrict ourselves to only members of the same species. On the other hand, if we just use a general description of what a "generic" tree looks like, we can more easily categorize a new example as being the same general category. Much more will be said about such categorization functions later.

There is some evidence, by the way, that people use this kind of general-feature interpretation to store and process images, but there is also evidence that people store and process with the actual pixel images (or perhaps some equivalent of the pixel images). Virtually everyone reports being able to recall and think about actual images, and psychological experiments indicate that they probably do. However, it is not likely that such images take the place of a more representational scheme; instead, it is more likely that they supplement it in some way. At this time, no one really understands how and why this should be so.

We have spent a great deal of time exploring the human visual system, and could justifiably spend much more. There are many thick books written about it; it is one of the areas of the brain that has received the most attention from researchers. Rather than delve into this in any greater detail, we should instead consider how what we know about vision in animals and people affects the construction of an artificial vision system.

As I implied earlier, most good artificial vision systems today mimic the human system, if only in that they perform so much processing in parallel. Few today try to build a vision system that runs serially; it is recognized that this is a losing proposition. In addition to mimicking the human visual system's parallelism, however, some excellent results have been achieved by following the kind of early processing-late processing example of the human vision system. One of the key challenges in vision is to make a system that can recognize the variety of objects that a person can, and that can do so as fast as a person. For example, the average person can recognize a simple picture of a house in about a tenth of a second. If it is a picture of a specific house, it may take a bit longer; if it is the person's own house or some other highly familiar house, it may take a bit less time. In any event, the idea is to achieve approximately this level of facility.

A major issue in achieving this level of capability is to recognize objects that are not in a known, fixed position or orientation. Two images of a house, for example, may be taken from slightly different positions, so that the house in one frame does not exactly match, pixel for pixel, the house in the other frame. Yet people have little or no difficulty in recognizing that both frames are pictures of the same object. How can this be achieved?

The most usual method of accomplishing such invariance is to transform the image before presenting it to the image-recognition system. And the most common method of doing this transformation is the Fourier transform, named after Jean Baptiste Joseph Fourier, the French mathematician who developed it.

The Fourier transform takes advantage of a peculiar characteristic of curved lines. A curve, even an irregularly shaped one, can be expressed as the sum of a series of mathematical sine- and cosine-terms. The reason this can be a useful expression of the curve is twofold: First, the original curve may not have a known equation that expresses it; second, sines and cosines are simple mathematical forms to work with, so expressing an arbitrary waveform in this manner may make it easier to deal with the curve.

A Fourier transform accomplishes another important objective, however. It converts a series of measurements—such as the list of bright and dark pixels in an image—from a spatial pattern to a frequency pattern. The transform is expressed as a collection of coefficients for a sum of sines and cosines of various frequencies. The critical information resulting from the transformation is the collection of coefficients for each frequency; these coefficients uniquely identify the transformed input pattern and collectively constitute the Fourier transform of the original pattern.

Fourier-transformations are often very useful in image processing for a couple of reasons. First, a Fourier-transformed image is very tolerant of translational errors in the image. In other words, a picture of a sailboat at the center of the frame yields much the same Fourier transform as a sailboat positioned along one side of the image. As a result, using Fourier-transformed images often makes object-recognition tasks much easier. Fourier-transformed images are also not very sensitive to rotational errors. Again, this reduces the amount of effort required to interpret the image because the Fourier transform doesn't change just because the object in the image is at an angle. And by using ring-cuts, described below, rotational effects can be almost completely eliminated.

A key problem in processing images is usually the sheer size of the pattern. An ordinary video camera image typically consists of 512 × 512 pixels, or more than a quarter of a million "dots"; a high-resolution image can have one to four million pixels. This is a huge number of individual data items to process. Using a Fourier transform does not directly reduce the number of data points, but a second set of procedures that operate on the transformed image can accomplish this. A two-dimensional Fourier transform of a typical image looks much like a series of dots scattered in concentric rings about a central point. Intensities of the dots at the center of the image correspond to low-frequency coefficients in the transformation; intensities of dots toward the edge of the image correspond to high-frequency coefficients; Frequencies themselves are represented by the distance of the dots from the center of the transform. For ordinary images, the low-frequency coefficients are typically much larger than high-frequency coefficients. Furthermore, the information contained in the dots can be consolidated by doing "wedge-cuts" and "ring-cuts."

A wedge-cut considers a triangular slice of the transformed image, much like the shape of a piece of pie. The intensities of all the dots within that wedge of the image are added together to generate a single number that represents the total intensity within that wedge (Figure 2.4). Typically, 8 to 16 equal-sized wedges might be used across the entire transformed image. The interesting characteristic of wedge-cuts is that they ignore size factors in an image. By using this technique, it doesn't matter much whether the object is large or small, the wedge-cuts give approximately equal values in either case.

A ring-cut considers concentric rings around the center of the Fourier-transformed image (see Figure 2.5). In the transformed image, low frequencies are at the center of the image and high frequencies at the outer edge. Thus, taking ring-cuts selects bands of frequencies to be consolidated together. Since in most ordinary images most of the dots are clustered at the center of the image in the low frequency range, there are two separate ways of doing a ring-cut. The first way uses equal-sized frequency bandwidths; the other uses approximately equal intensities within a band. In the first case, shown in

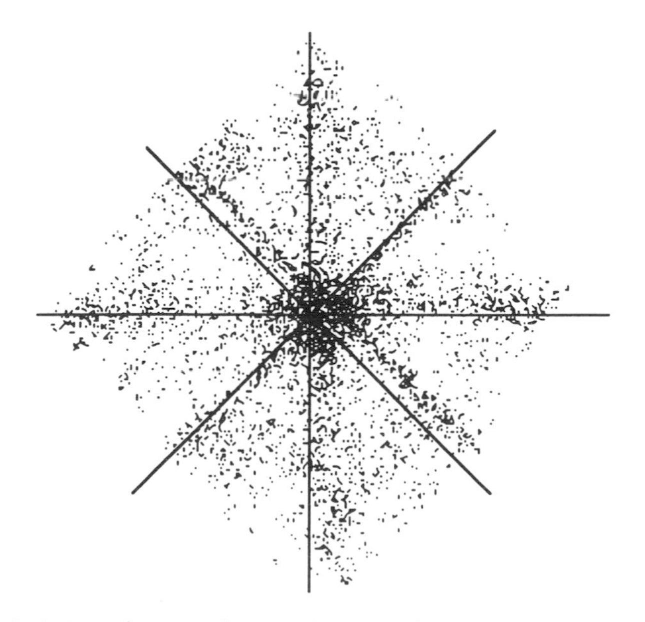

Figure 2.4 A wedge-cut of a Fourier transform. This example shows eight pie-shaped wedges. The intensities of the dots inside the wedges are summed to generate eight numbers that represent the entire transformed image.

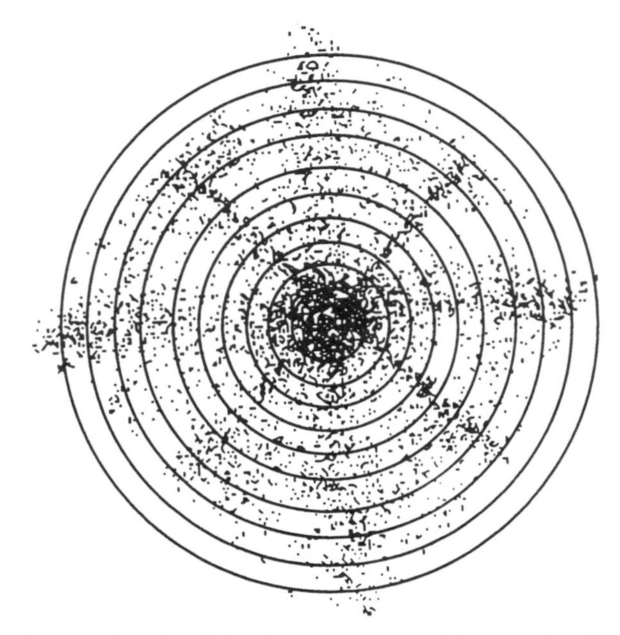

Figure 2.5 The rings in this ring-cut are equally spaced out from the center of the image. (The centermost ring is difficult to see because of the high intensity of the central dots in the image.) Intensities are summed within each ring; typically, total intensities in the innermost, low-frequency rings are much higher than in the outermost, high-frequency rings.

Figure 2.5, while the frequency range of each band is approximately equal, the sum of the intensities within the band vary sharply, with higher values in the low-frequency bands and lower values in the high-frequency bands. In the second case where intensities are held approximately equal, the bandwidths of the low-frequency bands are much narrower than those of the high-frequency bands, as in Figure 2.6.

For best results in processing images, it is usually better to include both types of ring-cuts. In general, ring-cuts provide additional rotational invariance to the Fourier transform. Just as with the wedge-cuts, the sum of the intensities of the dots within each ring is used as a characteristic for that ring. The number of rings needed depends on the problem, of course, but very good results have been obtained in image applications with 16 ring-cuts of each type, combined with 16 wedge-cuts.

In addition to providing some exceedingly useful invariance characteristics, the effect of using ring-cuts and wedge-cuts is to dramatically reduce the total dimensionality of the image. An image with a quarter of a million pixels can be reduced to fewer than 50 numbers (using 16 wedge-cuts, 16 equal-width ring-cuts, and 16 equal-intensity ring-cuts); obviously it's much easier to analyze the 50 numbers than it is to deal with a quarter of a million numbers!

Fourier transformations have drawbacks, of course, as well as advantages. The first of these is that actually doing the computations required to perform the transform itself is not an easy task; it takes a considerable amount of computational power to apply the transformation to a large array of values such as those found in an image. For best results, these are usually done on high-speed array processors, parallel computers, or even with special-purpose chips. A second drawback is that the process of converting the image from a spatial domain to a frequency domain discards some information contained in the image; sometimes this information is critical to the analysis of the image.

For example, consider a simple sine wave. A Fourier transform of that curve yields a single frequency (since a sine wave can be perfectly represented by a single sine wave!) You might think that the Fourier transform is a perfect representation of this curve—even that it ought to be able to perfectly represent this kind of curve. In actuality, how-

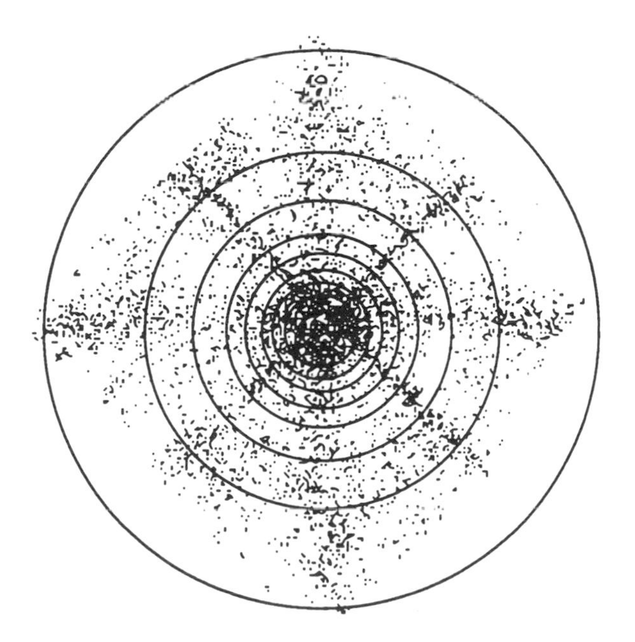

Figure 2.6 Another kind of ring-cut keeps the intensities approximately equal in each ring. This means that the outermost rings are much wider than the innermost ones.

ever, it cannot. The Fourier transform has completely lost all information about the relative phase of the curve. Figure 2.7 illustrates this point. The curve shown in (a) of the figure is a simple sine wave with a Fourier transform as shown in (b). Because only one coefficient is needed to represent all frequencies present in the sine wave, the Fourier transform is a single spike at that frequency. Curve (c) is another simple sine wave; in fact, it is just the curve in (a) displaced by half a cycle. But the Fourier transform of (c) is identical to that of (a); using just Fourier transforms of the curves we cannot distinguish between them.

For many applications, the loss of such phase information doesn't matter much. One good example of this is a production-line inspection system that used Fourier-transformed images (that were also wedge-cut and circle-cut) as input to a neural network. The network was to decide if the object in the image had any of half a dozen or so faults; for example, if the product being inspected was a bottle of shampoo, the list of faults might include a missing label, a bottle only partially full, a missing cap, the bottle tipped onto its side, and so on.

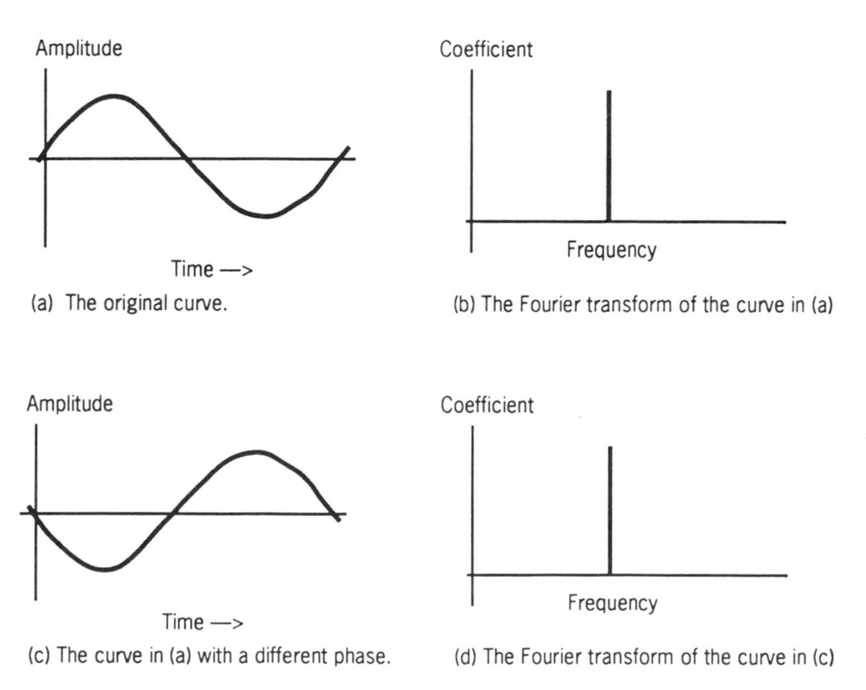

(a) The original curve.

(b) The Fourier transform of the curve in (a)

(c) The curve in (a) with a different phase.

(d) The Fourier transform of the curve in (c)

Figure 2.7 Fourier transforms lose all information about the phase of the original curve. Thus, the curves in (a) and (c) have identical transforms shown in (b) and (d).

The inspection system was so capable that it could achieve almost 100 percent accuracy, while inspecting up to 900 objects each minute. Clearly, the loss of phase information didn't matter much to that application. In other applications such as high-quality face-recognition, however, the loss of such phase data is devastating; Fourier transforms are ill suited for these cases and another preprocessing technique must be used.

While artificial vision systems have not reached human capability yet, dramatic improvements have occurred over the last decade. Some of this arises—and will continue to arise—solely because computers are orders of magnitude faster now than they were ten years ago and because parallel computers are now readily available, but were not then. But a great deal of progress has been made in the fundamental issues pertaining to how to build such a system as well.

One of the most successful of today's machine vision systems has been built by two German scientists, Ernst Dickmanns and Volker Graefe. Their system is novel because it deals with dynamic images, not just static "snapshots." The issue of dynamic vs. static images is an important one, because the real world is a very dynamic place. Environments change, objects move around, and a successful vision system must understand both the appearance of an object and its dynamics. Animal visual systems have relied on dynamical interpretations of scenes to such an extent that one of the most common defenses found in nature is for an animal to freeze its position. The animal's very lack of motion, usually combined with a camouflaging coloration, makes it literally seem to disappear into the background. Both predators and prey use this technique to escape detection.

Consider a simple scene with one object moving across the field of vision. If the object moves very fast, it is blurred and hard to distinguish details, but the motion itself is easy to detect. If the object moves slowly, it is relatively easy to fix attention on the object and thus details are more easily distinguished, but the motion may be harder to detect. In addition, if a moving object is of particular interest, an animal can fix its gaze on the object; similarly, a machine vision system can rotate its camera orientation to keep it in the center of the field of view. In both cases, animal and machine, the effect is to lose visual detail of everything except the moving object, which is increased in clarity. Interestingly, in animals the control of the physical motion of the visual system—the position of the eyeballs in their sockets, and the orientation of the head, for example—appears to be intimately linked to the visual system itself. The act of rotating the eyes or head to follow an object as it moves across the field of view provides some vital information about that object's relative or perceived position, distance, and velocity; this added information then assists the visual system in interpreting the scene.

The key to understanding dynamic scenes is the concept of continuity. A scene at one instant in time most likely is very similar to the same scene the previous instant. Thus, if a machine vision system can determine what objects are in the scene at some instant, it only has to deal with the differences between that "basic" scene and the one at the current instant in time. Since most objects in the frame will not change position or orientation much from instant to instant this dramatically reduces the difficulty involved in understanding the scene at each tick of the clock.

Consider the example illustrated in the three sections of Figure 2.8. Each section is a snapshot of a particular highway at three successive "ticks" of the clock. Can you tell what is happening in the images?

In the right-moving lane two cars are in the slow lane of the road, a convertible and a station wagon or van. The fast lane in this direction has a sedan that is in the process of passing the convertible, and rapidly overtaking the van. In the left-moving lanes, another convertible in the slow lane is rapidly being passed by a sports car. If we assume that the ticks represent equal time intervals, it is possible to form an estimate about relative speeds of all the vehicles, as well as their relative motions. For example, the sports car is clearly the fastest of them all, and the rightward-moving convertible is the slowest.

Now consider this sequence of frames from an image-processing perspective. Suppose that a machine vision system has identified all the objects in scene (a). If the system wants to locate the sportscar in scene (b), where should it look? The first place that the system should look is in the sportscar's location in scene (a), because the chances are that not all that much has changed in a single tick of the clock. Suppose the system draws a locator box around the sportscar's last known position that is about one and a half times the size of the sportscar in all dimensions. If the car's position hasn't changed much in one tick, it ought still to be somewhere in this box. Figure 2.9 shows where such a box would be; notice that the new position of the sportscar is indeed partially within this box. (If it were not, the box would have to be increased in size to accommodate the greater speed of the car.)

The point is that with this technique the machine vision system no longer has to search the entire image to find the sportscar; it only has to search the much smaller section of the image within the locator box. Obviously this is enormously cheaper in terms of computational

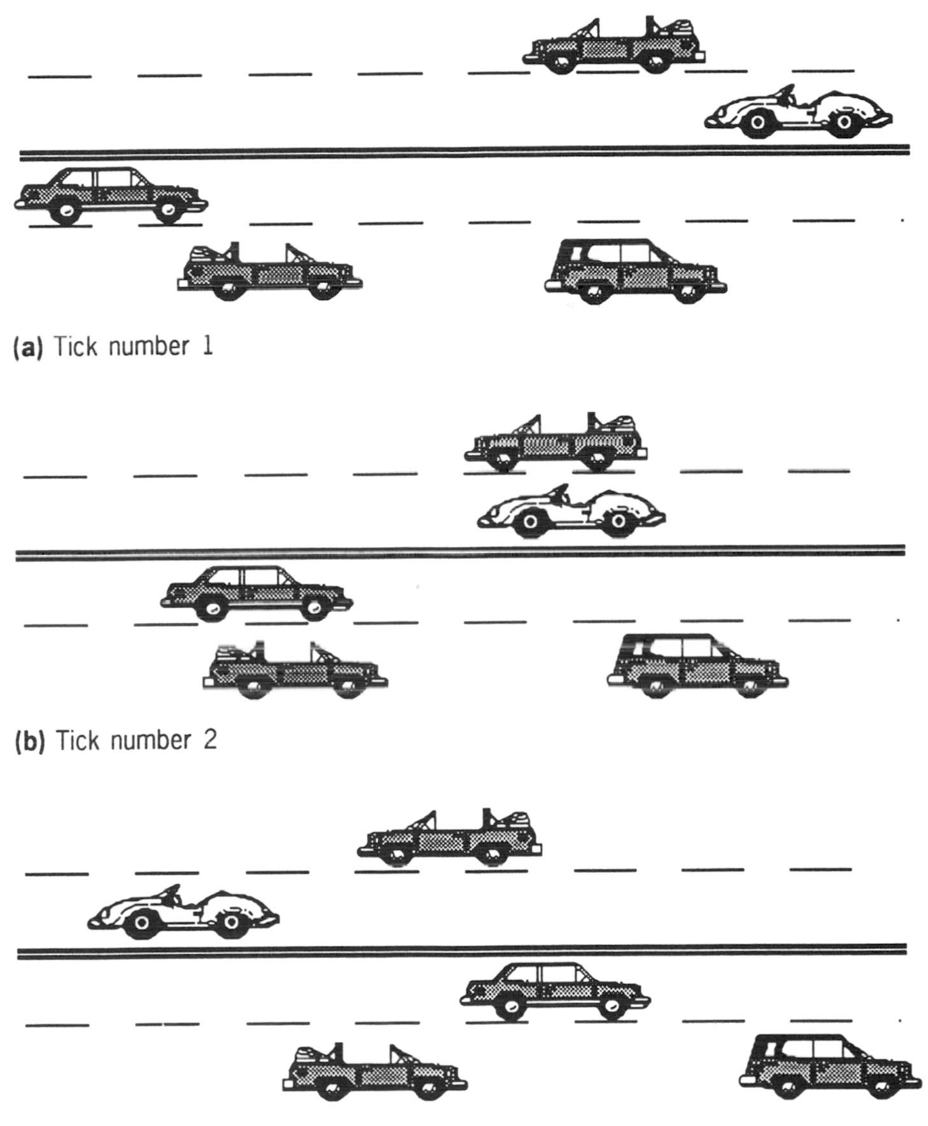

(a) Tick number 1

(b) Tick number 2

(c) Tick number 3

Figure 2.8 A freeway scene shown at three successive ticks of the clock.

time and effort than checking the entire image. In fact, most scenes are made up of details that are trivial and unimportant. It is not necessary that the complete image be fully processed; it is only necessary that the important parts of the image be understood. By taking advantage of the temporal continuity of objects in the real world—

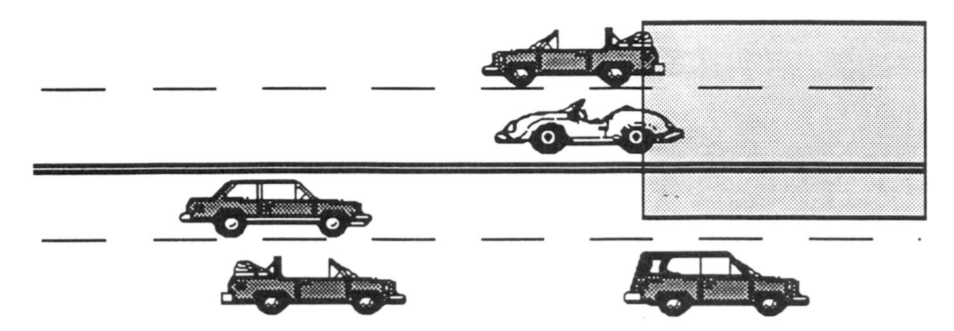

Figure 2.9 Tick 2 from Figure 2.8 with a box illustrating the most likely location of the sportscar in this tick.

that is, by making assumptions that physical objects don't appear and disappear randomly in time or space—the task of understanding complex, dynamic images is actually easier than that of fully understanding a single image.

As time progresses within a single scene, the ability to interpret and predict the locations of objects within the scene improves. For example, the locator box for the sportscar in tick 3 is likely to be longer, narrower, and extend more to the front than to the rear. Why? Because we now have some information about the car's current velocity, having observed it over two ticks instead of just one; just as position tends to be continuous in time, so does velocity. Very rarely do objects accelerate from a stopped position to a high velocity in less than a few ticks of the clock—and when they do, their motion is usually unobservable by humans without some kind of slow-motion camera.

Thus, as time passes the locator boxes are likely to get smaller and smaller as their predicted locations become more and more accurate. This means that less computational effort is necessary the longer the scene is observed. Furthermore, this frees up computer processing cycles so that background details or new objects that enter the scene can be handled. If the general scene persists in time, more and more background detail will be perceived; if not, little background detail is processed and understood. In other words, you really do see more of the countryside driving leisurely along a country lane than you do driving along a freeway—and the dynamic vision system experiences the same effect. In effect, the dynamic vision model takes advantage of the physical properties of the world to dramatically reduce the computational time and effort necessary to understand image sequences.

Another important concept of the Dickmanns and Graefe dynamic machine vision system is that the objects identified do not all have to have the same kinds of features. In other words, each locator box in the scene can be processed in an optimal way, depending only on the kind of object that is presumed to exist inside that box. For example, some objects may be easily defined by a series of edges in a particular orientation; locator boxes for those objects would be processed using edge-detection processing schemes. Other objects may be defined by varying textures, angles, or bars. In each case, only the minimal set of features required to identify the object and its position need be used in processing the appropriate locator box. Here again, a dramatic computational savings is achieved.

Such short cuts and time savers may seem to be subject to a lot of errors, and in fact they are. Two arguments counter this problem, however. The first is that people are not always correct in their interpretation of a scene; it is thus unfair and probably impractical—not to mention unnecessary—to expect a machine-vision system to achieve what people do not. The second point is that if the scene persists long enough, errors that occur in processing early ticks can be corrected with the computational time saved in later ticks. In other words, scenes that appear and disappear very quickly may well be perceived incorrectly in part; scenes that are present for substantial time periods, however, provide plenty of opportunities for correction and reinterpretation of the images. Again, this is similar to human vision performance.

Dickmanns and Graefe have applied their dynamic machine-vision system to a number of real-world control applications, ranging from balancing an inverted pendulum (the famous balance-a-broomstick-vertically-in-the-palm-of-your-hand trick) to control of an autonomously guided vehicle. Their results have been most impressive in all respects. A machine-vision guided car has been able to travel at speeds of up to 96 kilometers/hour (just under 60 miles/hour) under widely varying road conditions for distances of up to 20 kilometers (about 12 miles) with no human intervention. (They used real roads, both with and without lane markings, to test their system, with a human safety driver sitting in the front seat available to take control of the car if necessary; it was not necessary in their test runs.) Their system has also been able to perform lane changes and navigate highway entrance ramps safely.

We may not have androids that see as well as we do today, but excellent vision systems already exist in the laboratory. Furthermore, it is not necessary that an android's vision system match ours in capability and complexity. If the android can see (or sense by some other means) enough about its environment to be able to navigate and manipulate the objects it must to perform its assigned tasks, that will probably be good enough, at least for the first models.

We are learning more all the time about how our visual system works and how to imitate it in an artificial system. Someday soon we will be able to translate that understanding into our android children. The process has already begun.

≧ 3 ≦

Taking the First Step

Take most people, they're crazy about cars.
I'd rather have a goddamn horse.
A horse is at least *human*
for God's sake.

J. D. Salinger

In *Raiders of the Lost Ark,* hero Indiana Jones sets out on horseback to recover the Ark of the Covenant being trucked away by the evil Nazis. When asked how he plans to get the Ark back, Indy replies, "I don't know; I'm making this up as I go along."

Human beings are pretty flexible creatures, and highly adaptive to changing circumstances; nevertheless, Indy's plan in this case falls a little short of what one might hope to hear from the hero in such a tense situation. Luckily, Indy's famed resourcefulness comes to his assistance and all ends happily (except, perhaps, for the Nazis). Indy's problem in this case was at least partially one of locomotion; he had to figure out a way to get from where he was to where the Ark was as quickly as possible. This chapter considers how an android can deal with the problem of moving about in the real world.

Intelligent and successful creatures on earth typically have had two key characteristics: good sensory systems (usually with a high priority for vision), and mobility within their environment. As with all generalizations, there are exceptions to this statement, but in the main, intelligence appears to be linked with both the ability to perceive the environment and the ability to move about in it. On land this has led to the development of such animals as monkeys and apes, possessing superior color vision and high mobility, and in the sea, it has produced animals such as the dolphin and whales, which have intensely sensitive echolocation capabilities as well as the ability to roam the world's oceans. Mobility is so highly associated with intelligence, in fact, that an animal like an octopus, which can move around very freely, is about as smart as a cat or a dog, and much more

Figure 3.1 Not all robots have a manlike shape. Sometimes six legs are more appropriate than two for scrambling over rough terrain. (Photo supplied courtesy of Bruce Frisch.)

so than its cousin the sponge. (In contrast to the brainy octopus, the phrase "dumb as dirt" leaps to mind when describing the intelligence level of a sponge.)

Movement through the environment—locomotion—involves solving at least two separate problems, particularly when moving about on dry land. The first of these is the not-so-simple matter of maintaining a stable balance, both in standing and in walking. A child learns first of all to pull himself upright and stand. Initially, of course, the child holds on to a chair or a sofa to help maintain balance; only after practice can the child stand erect alone without falling. Once this is mastered, the child must then learn to maintain balance while moving about. This is not as simple as it seems, because walking involves a complex, highly synchronized side-to-side shift of the center of gravity of the body as weight is shifted from one leg to the other, all the while moving the "unweighted" leg and foot forward. Simultaneously, of course, the child must learn to control hip, knee, ankle, and foot muscles so that they lock properly into position to permit a stable stance, without becoming so tight that they prohibit motion completely. A young child just learning to perform this minor miracle of coordination and control is not called a "toddler" for nothing.

Human beings put a very high priority on the ability to move independently within their environment. An accident or disease that removes or hinders mobility is considered a major tragedy, and even a temporary loss of this function creates hardship, irritation, and emotional distress. While balance and walking are relatively straightforward, although complex, muscular control problems, navigational path planning requires the assistance of the vision system; thus a person who loses his or her sight must learn how to navigate all over again, even through familiar rooms and environments.

Walking is a complex process. With a being that has two legs, it requires a great deal of coordination. The center of gravity of the body must move from side to side while each leg takes its turn supporting it. For example, when the right leg is moving forward, the body shifts slightly to the left, so that the center of gravity is more nearly directly over the left leg. Without this shift in balance, the muscles of the left leg would be strongly stressed to keep the body from falling over when the right leg was lifted. Similarly, when the right foot is finally planted, the body's mass shifts back to the right, so that the right leg can support its weight more easily while the left leg lifts and moves forward. These body shifts are not extreme—after all a person's legs are not very far to the right and left of the center of the body to begin with—but they are essential for a stable walking motion. Even in a being with four legs, the body mass shift must occur during walking. It turns out, in fact, that only when there are five or more legs can the center of the body remain stationary with respect to side-to-side motion.

Robotic systems have been developed with as few as one leg, which achieves stable motion similar to that found by hopping on a pogo stick (and just as continuous as with a pogo stick, since if the robot stops hopping it falls over), and with as many as six or eight legs. Bipedal motion is the ideal in an android—remember, we are trying for a humanlike shape—but of all the requirements listed in Chapter 1, it is the easiest one on which to compromise. Almost certainly the first androids will either use wheels for motion, which will severely limit their mobility over rough terrain and along staircases, or they will have more than two legs because the coordination problems inherent in attempting to implement a bipedal stride are more difficult to overcome than those in a multilegged gait. Nevertheless, it is also likely that bipedal androids will follow shortly thereafter.

Let's consider just one walking robot system, one based on an evolutionary finite state machine approach. But first, we need to understand what a finite state machine is.

Figure 3.2 Attila and other "spider" robots have been developed by Rodney Brooks of MIT. Controlling and coordinating all those legs is easier than it might seem. (*Photo courtesy of Rodney Brooks.*)

A finite state machine is any device that has a specific number—a *finite* number—of possible conditions. The best everyday example of a finite state machine is a traffic light. A simple traffic light can be in any one of four possible states: it can have all lights off (i.e., it is broken or receiving no power); it can have its green light lit; it can have its yellow light lit; or it can have its red light lit. There are no other possible states for that traffic light. (Modern traffic lights often have additional states, such as blinking-red, blinking-yellow, and so on, but four suffice for this discussion.) It never, for example, has all three lights on, nor does it have its green light half on, and the yellow light three-quarters on. Each of the three lights is either on or off, and each of the states is well defined and predictable. Furthermore well-defined transitions exist between the states. When the light is in state green, a prompting stimulus can only cause it to change to state yellow. When it is in state yellow, a similar stimulus causes it to change to state red. And when it is in state red, another stimulus can cause it to change to state green. The lack of all stimuli (loss of power, for example) causes it to change to state all-off, no matter which of the other states it might be in.

Any finite state machine works just like this simple traffic light. It must have a limited number of states or conditions in one of which it exists at all times, and there must be specific stimuli that cause the machine to transition from state to state, just as the electrical pulses cause the traffic light to change from red to green. Sometimes there might be more than one possible path that leads to a specific state. For example, a few older traffic lights used to change from green to yellow to red to yellow to green. There were two possible ways the state yellow-on could be reached, one from the state green-on and one from the state red-on.

Usually finite state machines are diagrammed as a collection of circles, one for each possible state of the machine. The circles are connected by lines or arcs that represent the possible transition paths from state to state. The finite state diagram for the traffic light is shown in Figure 3.3.

Finite state machines are very useful indeed in computer science. A large body of research has gone into studying what such devices can accomplish. They are frequently used as a model for language understanding, for example, because the context of the first words in a sentence helps determine the meaning of later words. Thus, transitions from states such as "this is an unknown part of the sentence" to states like "this is a prepositional phrase" limit the parsing possibilities and reduce the time needed to process the rest of the sentence.

But right now we are interested in teaching a robot to walk, not in making it understand speech and language. How can finite state machines help create a walking robot? Rodney Brooks and other re-

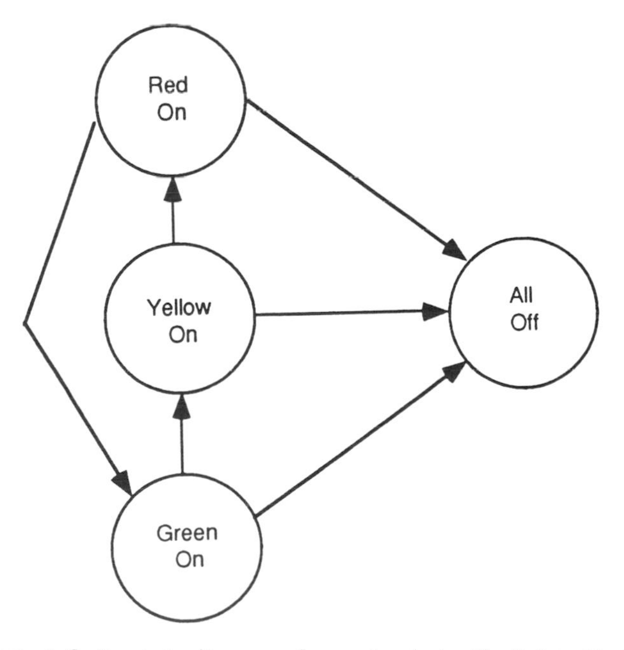

Figure 3.3 A finite-state diagram for a simple traffic light. Circles represent the possible states of the traffic light and arrows represent transitions between states.

searchers at MIT's AI Laboratory have developed a walking robot based on an evolutionary development approach, and that uses finite state machines as its model. This robot is a remarkable achievement. It has six legs and looks much like a giant metallic ant. Overall, the device is a little more than a foot long, has about a 10-inch "footspan" from side to side, and weighs a couple of pounds. It is entirely self-contained, and runs solely from batteries carried on its central body. Each of its six legs can move in two directions: up and down, and forward and back. The legs have no knee or ankle, only a single "hip" joint. At the front of the robot are two long "whisker" sensors that warn it of impending collisions with objects. It also has balance sensors that provide information on whether the body is inclined from the horizontal in either of two directions. Finally, the robot also has infrared "eyes" that can observe about a 45-degree arc in front of the robot. (The device, named Attila, is shown in Figures 3.1 and 3.2.)

This little beast is not a particularly intelligent device. It has no central computer telling it what to do, and its total memory capacity is only about 11,000 bytes—about enough memory to store 5 pages of text from this book. In contrast, if we make the over-simple assumption that each neuron in the human brain can hold only one character of information, the human brain's capacity is about 100,000,000,000 bytes, or about 50,000,000 pages of text—and this is far short of the brain's actual capacity. Even a fly's brain has about 1,000,000 neurons, giving it vastly greater capacity than this little robot. And yet this mechanical creature has some remarkable capabilities. To begin with, it can stand and walk while maintaining excellent balance. Furthermore, it can do so even when the terrain it is walking over is rough and difficult to navigate. With its whiskers it can anticipate when it is about to run into something, and it can veer to avoid a collision.

What is even more remarkable is that with its infrared sensors, the device only bothers to walk when there is something interesting nearby—like a person. When someone approaches the machine, its infrared sensors detect the person's body heat and cause the device to move toward the source of the heat. If the person stays in range of the sensors, the robot follows that person wherever the person goes. It is eerily similar to the imprinting of a baby bird just after hatching, when the hatchling follows the first moving object it sees, thinking the object is its mother.

This robot demonstrates several important points. It is constructed of a large number of small finite state machines, each of which performs a single, simple function. There is no central controller or brain, and none of the individual finite state machines is particularly intelligent about walking. Instead they perform functions such as raising or lowering a single foot, or moving one foot forward or back. In spite of this lack of central control, the robot is able to execute highly coordinated walking behaviors.

Furthermore, the sensory information from the whiskers and from the infrared eyes is not directly relayed to the legs. Instead, other finite state machines provide the walking mechanism with indirect information about the environment, and leave the problem of whether to raise or lower an individual foot to the specific finite state controller involved. The control of the system is thoroughly decentralized, rather than collected in a single location.

Finally, this robot was developed according to strict evolutionary principles. While we will consider global evolutionary issues in much more detail in Chapter 12, it is important to note that this system is consistent with an evolutionary approach. In other words, each subsystem added to the robot was added independently of all pre-existing systems. When the whisker collision-detector was added, for example, no changes at all were made to the basic balance and walking mechanisms. In spite of this, the new systems work beautifully with the older ones, allowing coordinated motion and more sophisticated behavior patterns. This is entirely consistent with what we know of the development of animal capabilities, and, as with most other aspects of the android, consistency with natural processes is likely to be a strong indication that we are on the right track.

The walking robot also demonstrates that the mere ability to move about in a coordinated fashion requires very little brainpower. The walking ability of this robot is not all that far short of that of an ant or a centipede. It can coordinate its leg motions to move around its environment; it can walk over obstacles and uneven terrain; is can even follow its senory detectors to move to desirable places. All these functions are strongly reminiscent of the locomotion capabilities of insects, spiders, and similar animals. This does not imply, of course, that the robot is as smart as an ant—ants have other behaviors as well as these locomotion skills. Nevertheless, it is difficult not to conclude that artificial systems are beginning to be extremely good mimics of some animal behaviors. The low level of computational effort needed to reproduce this walking behavior is also reassuring: If *every* major robot subsystem required the capabilities of a supercomputer, an android would be an unlikely development for another half-century or more.

It is entirely possible that the first working androids will have walking mechanisms very similar to those in this little robot. Walking is a much more useful method of locomotion than rolling in the real world. Few natural environments are smooth enough to make wheels a practical method of travel. Any real-world android will have to deal with stairs, curbs, rough terrain, and even escalators and rotating doors, just as people deal with these obstacles in today's world. It seems clear, though, that walking is an art that an android can master, even if it entails coordinating multiple sets of legs. There is no reason not to assume that walking—even bipedal walking—will be a skill that androids can certainly possess.

Getting an android to walk without falling over is only half the battle, however. A second, more cerebral problem exists before we can have one that walks around the house. This is the problem of navigating in the real world; it must be able to plan and follow a path from point to point. This means, of course, that the path must be a reasonable one to take; it must be relatively free of obstacles and fairly efficient. Path planning sounds quite simple until one tries to create a system that can do it. Think of the crush of people on any urban sidewalk at lunchtime. No matter how crowded the sidewalk may be, only rarely does one person actually physically contact another. An elaborate dance is performed that somehow successfully allows dozens or hundreds of people to move independently in a very limited area with remarkably few collisions. Each person must independently navigate through the crush of people, constantly updating and adjusting the direction of motion relative to those people nearby. Further, other obstacles and information must also be considered, such as the state of the walk-don't walk light at the corner, and the motions of the cars trying to enter or go through the intersection. Yet even in much less complex situations, the ability to successfully plan a path with no—or at least a minimum of—collisions with other objects is essential for successful navigation.

Consider the relatively simple problem of walking through a typical living room. Nearly always obstacles must be avoided, ranging from a low coffee table to this evening's paper on the floor, to the cat curled up on the rug in front of the fire. A successful path from, say, the door to the sofa must involve careful planning to avoid bumping into or stepping on these various obstacles. The furniture is generally not hard to miss, if only because it is usually arranged so that there are clear pathways to all seats in the room. If the room is a familiar one, a person can easily learn the appropriate path to follow. But what about the unexpected obstacle of the newspaper, just tossed onto the floor? Or even worse, the cat, which is itself animated and likely at any time to get up and move about the room?

In a laboratory setting, an android might be able to learn a collection of "hard-wired" paths to follow in order to move between known locations. But in the real world, those paths are likely to be strewn with changing conditions, obstacles, and other moving objects, all of which must be coped with properly. This means that the android will not be likely to encounter exactly the same conditions twice. Path planning must therefore occur on a real-time basis whenever the android is to move about. Otherwise, it can only blunder about in a very clumsy fashion.

In the last few years some tremendous strides have been made in developing systems that have autonomous navigational capabilities. The Dickmanns and Graefe self-driving car mentioned in Chapter 2 is one highly successful example of a device that can accurately and safely follow a roadway at high speed. But just following a given path, even at highway speeds, is not the same as planning a route through an obstacle course. The difference is that between following a given route to a destination, and choosing the route to follow. Following a given path may appear to be somewhat easier than the task of constructing the proposed path in the first place. In fact, however, path-planning systems were developed first, and only in the last two to three years have the autonomous vehicles that can follow those paths become available in the laboratory.

Constructing a path is a planning problem; it consists of much more than simply specifying a sequence of steps to follow to achieve a goal, however. In the real world, actions don't always have the consequences we expect, so any plan has to be able to both predict the results of each step, and also adapt to changes that arise from unexpected results of the plan. For example, suppose I construct a plan to bake a cake. One step of the plan is to add three eggs to the batter. Now, I cannot just dump in the eggs; I must first crack the shells carefully so that the white and yolk drop into the bowl, while at the same time, not allowing any small pieces of shell to fall in as well. I might construct a plan that looks like this:

1. Open the refrigerator door and remove the carton of eggs.
2. Shut the refrigerator door.
3. Place the egg carton on the counter, open it, and remove three eggs.
4. Place the three eggs in a small bowl (so they won't roll off the counter), and close the carton.
5. Open the refrigerator door, pick up the carton, and replace it in the refrigerator.
6. Shut the refrigerator door.
7. Pick up one egg from the small bowl on the counter. Holding it carefully, tap it firmly once against the corner of the counter, so that a crack appears in the shell.
8. Using two hands, move the egg so that it is over the bowl of batter. Carefully pull the two halves of the shell apart so that the egg runs into the bowl.

9. When the egg shell is empty, place the shells in the sink for later disposal.
10. Repeat steps 7, 8, and 9 with each of the remaining two eggs in the small bowl.

Now this plan looks pretty comprehensive, but several snags might appear. For example, suppose there is no carton of eggs in the refrigerator when the refrigerator door is opened? Or suppose that when the carton is opened there are only two eggs inside? Clearly, there should be a step in the overall bake-a-cake plan that checks to see if there are sufficient ingredients before the cake-making process is actually begun.

But there are other potential difficulties as well. What if there are exactly three eggs in the carton? Once they are removed from the carton to be added to the cake, should I then replace the now-empty carton in the refrigerator? Of course not; thus the plan should have a prediction attached to step three that expects to see at least one egg remaining in the carton; if there are none left, the plan must have a contingency action of, say, adding eggs to the shopping list on the refrigerator and discarding the carton in the trash.

Suppose when I tap the egg against the edge of the counter, the egg is tougher-shelled than usual and does not crack. The plan should also have a contingency to deal with this. Similarly, if I tap it too hard against the counter so that the egg smashes completely, it should have a contingency plan to clean up the mess and get another egg out of the refrigerator. And what if one of the eggs is rotten? A better plan would crack each egg into a saucer for inspection before adding it to the cake batter. Then if an egg turns out to be rotten, it can be replaced with another without damage to the batter.

In each case, an intelligent plan should know what the desired result of each step should be, and should be able to regroup and deal with situations that don't go exactly as planned. This implies that path planning has an important element of prediction: In order to construct a plan, the results of each step—along with the potential reactions of other beings or devices—must be predicted. These predictions then allow the android to estimate the state of the world that results from the execution of each step of the plan, and thus decide on the following action. Planning requires a more careful strategy than Indiana Jones's make-it-up-as-he-goes-along "plan."

Planning programs can be sorted into a couple of categories. There are those that use only generalized techniques that work in a wide variety of situations, and those that are domain-specific, and which take advantage of special knowledge or heuristics specific to the particular kind of problem being solved. Domain-specific planners tend to be more competent in specific circumstances because of their additional knowledge; on the other hand, generalized planners have the additional flexibility needed to deal with a broad range of problems.

In order to build a plan to do something, an android must first have a goal—a task to accomplish or a destination to reach. In addition, the system must have a representation of the current state of the world, as well as a prediction of the state of the world when the goal is achieved. ("World" in this context means only the android's current environment: those situations that can directly affect it or objects that are in the android's immediate proximity.) For example, if the goal is to walk across the room to the window, the current state of the world should include a sensory representation of the objects in the room, their positions, and their relative mobility (can the cat on the floor get up and move, for example). Included in this representation is the current position of the android, as well as its orientation: facing the door, facing the window or some other direction, upright or prone, and so on.

To construct a plan, the android also has to have a catalogue of actions that it can take to try to achieve it. For example, its catalogue might include picking something up, of dropping whatever it carries, of moving forward or sideways or backwards, and so on. Each of these actions has a possible consequence that will change the perceived state of the world. For instance, if the android moves forward three feet, its position within the room is now different. If it picks up a newspaper on the floor to clear a walkway, the world has changed yet again.

Construction of a plan in this sense merely means to build a sequential list of actions that the android needs to take to achieve its goal. This implies that it has to project the probable state of the world after each action and then use that altered state to determine the next action to be taken. When the projected result of one of these actions matches the desired goal, the plan is complete.

Planning is frequently considered a search function. This means that the collection of all possible sequences of actions are searched to find one (not necessarily the only one, or even the best one) that successfully reaches the goal. If such a sequence is found, the goal is achievable; if not, the goal is impossible. The more actions the android can perform, and the more distant the goal, the larger this search space is. This has two consequences. If it has many possible actions, there may be a lot of sequences that result in success; on the other hand, this also means that there are more possibilities to consider at each step in the plan and thus more processing is needed at each step. Also, if the goal is very far away, the number of possible paths that end up at the goal state may be enormous. In this case it is likely that some of these paths will be very inefficient compared to others. Under these circumstances, it is often necessary to perform a secondary search among all derived successful paths in order to determine one that is relatively efficient. This is called optimizing a plan; a classic example is the Traveling Salesman Problem of finding the most efficient path that passes through a large number of cities without visiting any city twice.

Plans can be constructed in two basic ways. The first and most common method is to have the system develop the complete plan before executing any step of the plan. Take the example mentioned earlier of moving from the door to the window. An android attempting this task might formulate the following plan, illustrated in Figure 3.4.

1. <BEGIN PLAN> Move forward three feet.
2. Pick up the newspaper from the floor.
3. Fold the newspaper.
4. Move right 2 feet.
5. Put the newspaper on the coffee table.
6. Move forward 2 feet.
7. Move left 3 feet to avoid the cat.
8. Move forward 2 feet.
9. Move right 4 feet.
10. Move forward 6 feet to window. <GOAL ACHIEVED>

This looks like a simple method to get to the window. In formulating the plan the android uses only a few basic operators: move forward, move left, move right, pick up an object, put down an object, and fold an object. Some of these actions—fold an object, for example—may themselves be calls to other plans that tell the android exactly how to perform them. For the purposes of this plan, however, all these actions can be considered fundamental actions that it already knows how to perform.

To develop this plan in its entirety demands that the android recognize various objects in the room: the coffee table, the newspaper, the cat, the window. It also requires that the system know what actions are appropriate for each object; for example, it would not be acceptable for the android to pick up and fold the cat in an attempt to straighten the room—the cat would most certainly object to such cavalier treatment. It is also true that the plan achieves a subgoal that may not be apparent: It properly disposes of the newspaper on the floor, even though that subgoal has nothing to do with the main goal of getting to the window. The android might have

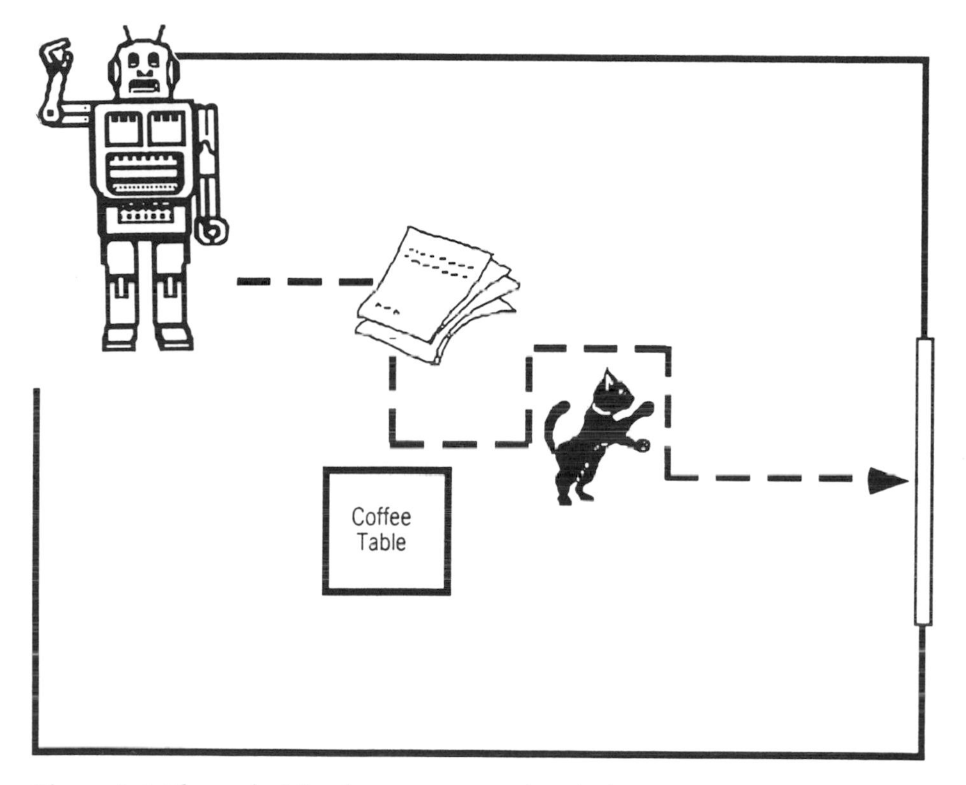

Figure 3.4 The android's plan to move to the window.

simply stepped over—or on!—the paper on its way to the window; it had no need to actually pick it up and put it away. Apparently this android has some ever-present goals that it always tries to satisfy, including something to the effect that whenever it encounters an object that is out of place it should relocate the object properly.

Any number of planning systems constructed in the last ten or fifteen years would be able to develop and execute this plan. The key element of planning is that a plan carries no commitment to execute any of its steps. In the simple case here, the order in which the steps are carried out is vital; step 3 cannot possibly be carried out until step 2 is complete, and so on. Even the motion sequences are in a critical order. To see this for yourself, try the following experiment. Hold your right arm straight out in front of you. Now execute the following two-step plan:

1. Rotate your arm straight up.
2. Rotate your arm to the right.

Your arm ends up pointed directly to the right. Now extend your arm directly in front again and execute the same plan, but in a different sequence:

1. Rotate your arm to the right.
2. Rotate your arm straight up.

Where does your arm point now? The same plan executed in reverse order yields an entirely different result. Typically, directional motion plans have this characteristic.

The plan outlined above for the android to move to the window is barely adequate. For example, what if the cat moves while the android is executing the first few steps of its plan? If the cat gets up and moves out of the room, or jumps on top of the coffee table, why should the android do the little zigzag of steps 7–9? Yet if the plan is fully developed before the android moves, and if there is no process for it to update the plan during execution, the android will do its sidestep even if the cat is no longer present as an obstacle. Planning systems that perform this kind of complete analysis in advance must provide some mechanism to alter the remaining steps in the plan if the world changes in unexpected ways.

Modern planning systems can construct a much more sophisticated solution to this problem. In these planners, the android postpones deciding what to do until it really has to make a decision. The complete plan is not prepared all at once, as it is above. Instead, only enough of the plan is generated to be sure of a reasonable first step (or enough steps to reach another major decision-point). Thus, the android might decide that it should move in the general direction of the window—forward—until it reaches some kind of obstacle. When stopped at the newspaper, the android would at that time decide what to do next. In many ways, this has much the same flavor as Indiana Jones's making it up as he goes along! And while the actions that the android actually takes may not differ at all from those listed in the 10-step plan above, the very fact that these decisions are made on the fly makes this approach more flexible than the previous one.

The advantage of preparing the complete plan all at once is that the android does nothing until it is assured that a procedure exists to actually reach the desired goal. If, for example, some uncrossable barrier exists between the android and the window, the android does not even attempt to get there and returns a response that says, in effect, "there's no way to get there from here." (This mimics the conclusions arrived at by many tourists trying to navigate through Boston's fiendishly laid-out one-way street system.) On the other hand, planners that do this complete an analysis of a problem often have the disadvantage that they are either inflexible when encountering unexpected changes in the world environment, or they require an unacceptably high level of computational effort to prepare the plan since they have to account for all kinds of intermediate world-states.

The advantages carried by the plan-as-you-go approach are, of course, flexibility and speed. Since only the most immediate next steps have to be determined, the system can act much more quickly than if it tried to decide on a complete plan before doing anything. This also allows the plan to be much more flexible. At each decision-point the ensuing steps are based on the actual results of previous actions, not on estimates or guesses of what might happen. The disadvantage of this is that there is no guarantee that the goal is ultimately attainable. It is entirely possible that such seat-of-the-pants guidance may not be able to reach the goal state at all.

The plan-as-you-go approach has many characteristics of a "greedy" algorithm. A greedy algorithm is a step-by-step solution to a problem in which, at each step, the action taken is the one that seems optimal at that step. There are flaws with this approach of course, most notably when a short-term loss must be taken to achieve a longer-term benefit or goal. Nevertheless, such greedy solutions work very well for many problems in the real world.

The determination of which approach is best depends largely on the domain of problems the android is expected to deal with. In many cases it can be expected that real-world problems will have attainable goals and the plan-as-you-go approach works fine. For some, more subtle cases, however, it may be necessary to resort to complete pre-planning to assure that the goal is achieved. It is probable that a working android will have both kinds of planning systems in place, so that it can choose the best technique for each situation based on its prior experience with this or similar problems.

It seems likely that locomotion, navigation, and balance systems for an android are well within our grasp even today. The key limitation of planning systems is the high level of computational resources required to solve complex problems. As computers get more powerful, smaller, cheaper, and faster this limitation is certain to become less stringent. With regard to mobility, the android can be limited initially by giving it wheels instead of legs, but with the tremendously successful multilegged systems that have already been developed, this also is likely to be well resolved in the next few years. While there are still some important problems remaining to be solved, locomotion and navigation systems will almost certainly be ready for the android by the time the android is ready for them.

≧ 4 ≦

An Android's Reach

We're more alike than you think.
If you prick me, do I not—leak?
Android Commander Data of
Star Trek: The Next Generation

Robert Browning wrote that "a man's reach should exceed his grasp, or what's a heaven for?" That may be true on a philosophical plane, but an android has more practical requirements. An android needs to manipulate the objects around it—or at least some of them anyway—and to do that implies the existence of a hand and an arm.

A human being has a quite wonderfully coordinated body. Even the clumsiest of us can manage to pick up an object from a table and move it to another desired location. We do not need to consciously think about the path the hand and arm takes from its initial position to the object, nor about the path from that position to the final resting place. If there are other objects in the way, a person will quite naturally reach around them, or move them out of the way, or change position to better reach the desired object. For example, if asked to pass the salt when eating at a diner, no one finds it particularly difficult to solve the problem of reaching around the catsup bottle to get to the salt shaker.

A simple robot arm, on the other hand, would have quite a hard time trying to figure out how to get to the salt shaker. Like the human arm, a robot arm typically has several joints that are equivalent to the shoulder, elbow, and wrist. (Some robot arms have even more joints, but these three suffice for this discussion.) For any given starting position, such as the arm at rest on the edge of the table—assuming no one has yet taught it to keep its elbows off the table!—to the desired finishing position with the hand grasping the salt shaker, there may be a huge number of possible paths the arm might travel.

Which path will allow the arm to reach the salt shaker without contorting into an impossible position?

Furthermore, the robot must not only determine what path to take, but also how to move the arm at each of the available joints to follow that path. Suppose the robot arm is to wave its hand in the very simple motion of a circular arc. How does it decide what angular changes are appropriate at each of the joints (shoulder, elbow, and wrist in this example) to make the circular arc desired? It turns out that the mathematics involved in solving such problems is highly complex and not all that easy to do. As a result, many robots today are not very good at complex coordination problems such as reaching for the salt shaker. In other words, compared to an even moderately talented person they are more than a bit clumsy.

Suppose that the salt shaker has been knocked over and is lying on its side. A person has no trouble at all in changing the grasping motion to accommodate the new orientation of the shaker; a robot typically finds this to be fairly difficult. And if the salt shaker is in a peculiar shape—perhaps that of a small plastic cow—there would likely be additional difficulties while the robot tried to pick it up.

Even if we suppose that the problem of coordination is solved, there is still another pitfall awaiting the android. Suppose that the request is not to pass the salt shaker, but to pass a feather instead. Can the android moderate the force it uses to pick up the feather, and thus avoid an embarrassing upward jerk? And what if it is told to pick up a heavy brick?

A robot arm tends to do pretty well picking up items that are of a particular known weight. However, if we load down the arm with an unexpectedly heavy object, like a brick, or (in some cases) an unexpectedly light object, like a feather, there is a tendency for robot arms to exhibit unpredictable behavior. The amount of force needed to pick up and move the light object is much less than the force the robot actually applies, for example, so the arm swings wildly. Alternatively, the force needed to pick up the brick is much more than expected, so either the brick doesn't move at all or, if it moves, the unexpected weight at the end of the arm again causes unpredictable arm motions. Yet here again, a small child can easily adapt the force exerted to that needed to accomplish the task (assuming the child can lift the heavy object at all, of course). Only if the object is much too heavy does the child exhibit truly uncontrolled motion, staggering around wildly during his or her efforts.

And this is not the entire problem either. If the object requested is a brick, it doesn't matter all that much how the robot handles it, as long as its grip is secure enough to manipulate it. But if the object is an egg or a delicate piece of crystal, too light a touch will allow the object to drop and break, and too heavy a grip will smash it. Children have to learn this fine motor control too, of course, which is one reason few parents trust their fine china to their young children's tender mercies. Still, people seem to do extraordinarily well at manipulation tasks that cover a wide range of force, from the lightest touch to the heaviest hand.

Robotic arm control is complicated still further by other issues. Suppose that the arm is ordered to rotate 10 degrees to the left. It does not necessarily follow that the actual physical motion that results is precisely 10 degrees; in fact, there is almost certainly some small error associated with any robot arm motion. Any number of physical influences can cause these errors, ranging from simple friction and gear backlash to room vibrations or inaccurate calibration of the equipment. Feedback that determines the actual arm position and motion —as opposed to the expected position and motion—becomes essential for a practical robot arm system. Otherwise, the small errors from each individual movement accumulate so that a complex motion may result in a wildly inaccurate final position. Even more problems accumulate in this regard if the robot is required to touch an object, since the rigidity of the surface touched strongly affects the arm's response to motion commands—and what good is a robot arm that never touches anything?

One solution to these positioning problems is to provide two separate controlling systems for the robot arm: a coarse control that has the primary duty of dealing with gross movements of the arm, and a fine control that performs fine motor control of the hand and wrist. Typically, a fine control that operates in a broad range of environments is designed as a multifingered hand with tactile sensors; one that is used exclusively in a constrained environment (such as on a manufacturing line) has a highly specialized design suited to that environment.

One of the reasons people do so well at grasping and manipulation tasks is that so much of the brain is dedicated to controlling the hand. Astonishingly, maps of the brain indicate that a huge amount of the cerebral cortex—that part of the brain involved in higher thinking—is involved in controlling just the hands, more than almost any other area of the body. Figures 4.1 and 4.2 illustrate the relative sizes of the portions of the cerebral cortex that control various parts of the body, with Figure 4.1 illustrating the amount of cortex used to process sensory information, and Figure 4.2 showing the amount of cortex used to control the same body parts. The entire torso is controlled by a fraction of the cortex used by the hand, for example. Apparently, controlling the hand is one of the brain's most important functions—only the mouth and vocal systems take a comparable amount of brainpower—so reaching and grasping functions are likely to be equally important for an android as well. Furthermore, based on our own brains, we can expect the robot to use a considerable portion of the android's total computational power to handle grasping and coordination problems.

Consider the problem of grasping an object. How is it that people can grasp and manipulate such a wide range of objects? The key to this must lie in the notion of feedback. The human hand is not merely something that manipulates objects, it is also a highly sensitive device for discovering information about objects. Touch is one obvious means of determining information about a given object, of course, and the human hand has highly sensitive tactile receptors in the fingertips. These touch receptors pass information about the texture and composition of an object back to the brain, which can then use this information to help determine how much force is necessary.

But this is not the entire story. The brain uses visual information as well as tactile data; an object that looks like a brick is expected to weigh like one also (or more precisely, have the mass of one); one that looks like a delicate piece of crystal is likely to have a similar lightness and delicacy. Of course, these combined sources of information are occasionally fooled by objects that don't look very heavy but are, or by other objects that appear to be massive but are really very lightweight. This leads to a third important means of measuring an object's properties: muscular feedback.

When we try to lift an object of unknown weight, often we first give it an experimental tug. The resistance the hand and arm muscles encounter in that tug gives us a great deal of information about the object's weight. In some people, this muscular feedback is so highly tuned that they can accurately determine the number of ounces in a letter just by considering how heavy it feels in the hand. But even when this ability is not particularly developed, it still is a primary source of information about objects we are currently manipulating. Such sensory feedback will overrule the visual and tactile information if necessary, enabling us to adjust the force needed to lift an object whose weight does not match its appearance. Thus a person has no trouble lifting a hollow brick; after a very brief time the initial "heavy-handed" effort is modified to be more in keeping with the force actually required.

As we have seen, people have highly efficient and effective systems for coordination and grasping. Do robots do as well? Unfortunately, for most cases the answer has to be no. Robot arms came into industrial use in the 1960s and great predictions were made at that time about what the factories of the future would look like. These robots generally were only able to repeat a specific physical motion;

they had little feedback or "intelligence" in the sense of moderating their actions to suit non-standard conditions. If they were to pick up a widget from a conveyer belt and put it in a particular place, the widget had to be exactly where it was supposed to be, and in the proper orientation. If the widget fell over, or was out of position, often the robot could not pick it up, or, on picking it up could not manipulate it properly to make the widget fit where it was supposed to go. In summary, they were very stupid robots indeed.

Many industrial robots today have changed little from these simple models. While there are some interesting cases of factory floors that are nearly completely automated, these remain the exception and not the rule. Today's industrial robot is often—though not always—nearly as unintelligent as its predecessors in terms of having

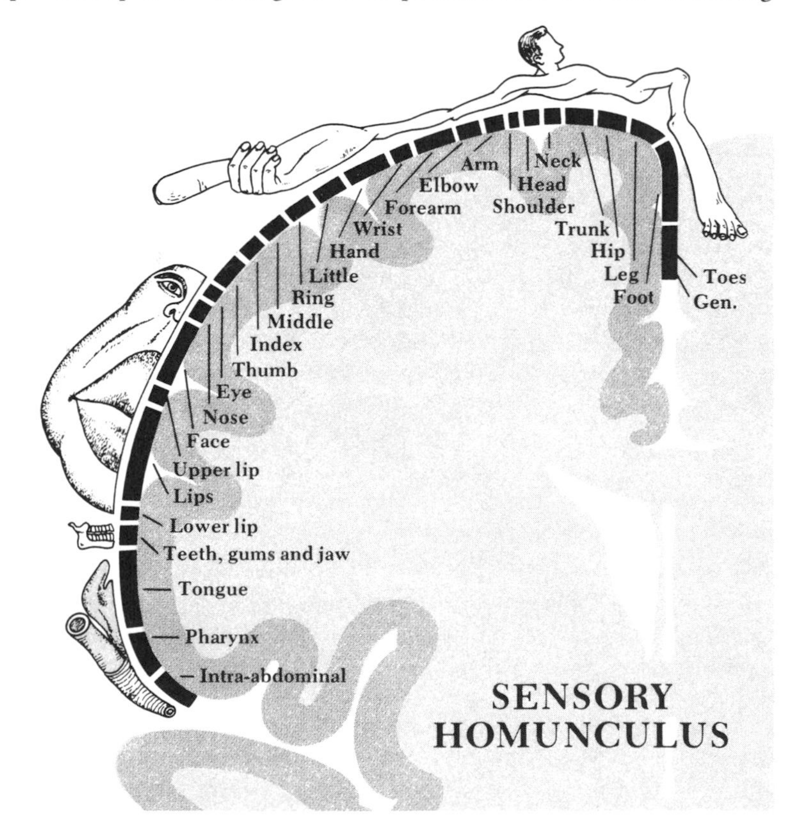

Figure 4.1 The human body as perceived by the sensory processing functions of the cerebral cortex. Relative sizes of the body parts indicate the amount of cortex used to process sensory data. (From "The Great Ravelled Knot," by George W. Gray. Copyright © 1948 by Scientific American, Inc. All rights reserved.)

reasonable common sense to perform simple tasks. This is changing, however, and the next generation of robots now beginning to make its appearance in the market offers dramatically improved performance.

The problem with arm path planning is that the world is not a neat and tidy place. If the goal is to pick up a particular object, very often there are other objects in the way that must be reached around. And generally we don't want such obstacles knocked over or damaged in the reaching process. This is a problem in eye-hand coordination. That is, the eye has to guide the hand so that it smoothly reaches the desired object. For years, researchers have tried to solve this problem through a reasoning approach. In some cases the results were very good, but the problem is that such a solution has to know how to

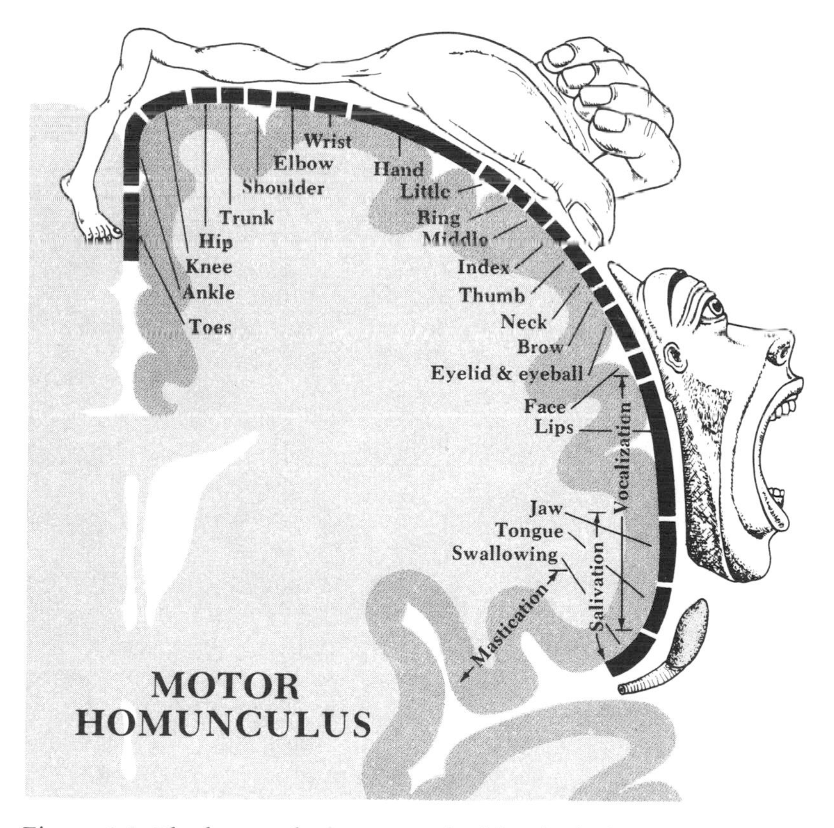

Figure 4.2 The human body as perceived by the body control functions of the cerebral cortex. Again, relative sizes indicate the amount of cerebral cortex devoted to processing information from each part of the body. (From "The Great Ravelled Knot," by George W. Gray. Copyright © 1948 by Scientific American, Inc. All rights reserved.)

solve nearly every possible combination of orientations and locations relative to fixed obstacles. And if the obstacles change position, this too must be added to the system. Pretty soon, the number of rules and examples grows to the point where the system is no longer a real-time system. The controlling computer simply can't compute fast enough to consider all the possibilities.

Mitsuo Kawato, Kazunori Furukawa, Hiroyuki Miyamoto, and Ryoji Suzuki of Osaka University, have developed a robotic arm that is far more sophisticated than most other robotic arms in use today. Their system is based on an analysis of the brain circuitry involved in human arm motion control. In a person, there is a complex collection of sensory feedback that helps the brain know how well the arm is carrying out a given task. In addition, of course, visual information can also be used to moderate arm control signals. In these researchers' view, the problem of, say, lifting an egg is carried out by a complicated series of control systems that interact highly with each other. For example, the initial command to "pick up the egg" is first translated into a collection of commands that serve as a rough guide to the task that needs to be done. The commands might indicate that the hand is to be moved in a particular general direction or toward (or away from) a particular object. This crude order is then transmitted to the section of the brain responsible for actually generating commands to the muscles, the motor cortex. The motor cortex then generates an equally crude set of muscular commands that are sent out to the arm muscles for implementation. These cause the arm to begin moving toward the desired object.

Since the commands are only grossly accurate, the initial motion is similarly likely to be correct in only a limited sense. Feedback must be added to make the system respond smoothly to the command. Of course one kind of feedback is vision, and this is very useful indeed. However, for many tasks, people simply don't use their eyes as a primary source of arm trajectory control. Anyone who has ever read a newspaper or book while eating knows that it is unnecessary to actually watch the fork or spoon in order to have it correctly move between the mouth and the plate. And any teenager will assert that he or she can watch television and still manage to do any number of complex mechanical processes—like write the homework that is due tomorrow! So while visual feedback is an important part of the Kawato system, more is needed for smooth operation. The Osaka team added two other kinds of feedback to their model.

One type of feedback arises from a system that constantly compares the roughly estimated path of the original command ("pick up the egg") to the actual path the arm is taking, derived from sensory input from the arm. This system has learned through experience what the likely outcome will be of any given motor control command to the arm. In essence, it acts as a supervisor, watching what happens when a given arm command is actually performed. It uses this experience to predict what the probable outcome of each command sent to the arm will be. This gets compared to the crudely expressed desired path, to generate signals that can constantly update and correct both the motor cortex and the arm muscles.

The second kind of feedback used by the Osaka system is one that monitors the internal workings of the system itself, rather than the actual performance of the arm. Essentially, it becomes an expert in knowing what kinds of motor commands are appropriate for any given desired task. Its usefulness is its ability to generalize from one task to another, so that new tasks are performed easily based on similar, known tasks. It also helps to smooth the arm's motion even when other circumstances are changing, as might happen, for example, if the arm is mounted on the moving platform of a mobile android.

Using this complex model, the robot arm developed by Kawato and his colleagues has been shown to be capable of learning general motion under varying conditions. The arm is also capable of handling a range of loads, although more complete load flexibility depends on adding some additional kinds of feedback subsystems to the arm. Their system can pick up objects in arbitrary positions, without having to be provided with an exact trajectory. This is a major improvement over earlier robotic arm systems.

Just outside of Boston is a small company called Neurogen. Begun by Michael Kuperstein in the 1980s, it has developed some innovations in robotics. One of its systems has achieved a similar breakthrough in the problem of reaching and grasping objects in unconstrained environments. The people at Neurogen took quite a nontraditional approach to the problem of determining robotic arm trajectories. Like the Osaka researchers, they first considered how a person learns to solve this problem. The key difference between Kuperstein's approach and Kawato's arm controller is that Kuperstein modeled the *development* of human eye-hand coordination, not the operation of a fully functional adult.

Babies randomly wave their arms around, only gradually connecting their deliberate muscular actions with the funny pink sausage-thing that moves in and out of their visual field. Gradually they make the connection that by controlling certain muscles they can make the sausage-thing move where they want. Still more experimentation is needed for the baby to get a handle on grasping objects. Finally, however, the baby makes the connection between what it sees and the muscular motions it makes. The first step toward a highly coordinated reaching and grasping system has been taken.

Given this basic example of how people learn eye-hand coordination, Neurogen set out to build a robot arm to achieve the same performance. Starting with a system containing two video cameras (for stereoscopic vision) and a single robot arm, they set up a complex self-organizing neural network,* complete with feedback. (The system also included a great deal of other processing equipment as well, particularly to handle the visual images.) The idea was to allow the network to go through the same learning period a baby does in which the connections between muscular control and visual input is gradually made. Neurogen's system, called INFANT for obvious reasons, does precisely this.

INFANT learns to map visual input patterns to muscular commands sent to its arm. It uses a self-organizing neural network to learn this mapping, and, just like a baby, it must experiment with its world in order to learn. In essence, it randomly "waves its arm" around and watches the resulting motion with its video camera eyes. As with a baby, learning is not instantaneous; INFANT must make many attempts before control of its arm is certain. Eventually, however, INFANT learns to coordinate what it sees with what its arm can do. The result is a robot arm that can pick up any object within its reach, no matter what the object's position or orientation. Unlike more conventional robots, INFANT needs no explicit programming to supply it with rules for reaching and grasping because it develops its own knowledge as it self-organizes.

While not a complete solution as yet—INFANT is still under development by Neurogen, along with yet more advanced robot systems—this system clearly demonstrates the kind of advances that can be expected in the next few months and years in the areas of coordinated robot arm systems.

But what about the robotic hand itself? Researchers in many laboratories are also extending the manipulation capabilities of the robot hand. Robots can now crack eggs, shear sheep, play Ping-Pong, and tighten wing-nuts, as well as perform other feats of dexterity. One robot, the WABOT-2 developed in the mid-1980s at Wasanaka University in Japan, can read simple sheet music and play it on an electric organ or synthesizer keyboard. It can also listen to a human accompanist and adjust its tempo to that of the human's performance. While it

*A self-organizing neural network is one in which the training is based solely on the input data received, rather than on the input data combined with known, desired outputs. In essence, a self-organizing network determines for itself what the desired output should be for a given input stimulus. Chapter 7 discusses these networks in more detail.

cannot handle complex tunes, it can play better than many humans, and certainly better than I can! As we can see by these brief examples, robot hand dexterity has already improved dramatically, and even now a number of advanced systems have near-human performance in specialized tasks. Several research laboratories expect to develop a general purpose robot hand that approximately matches human hand dexterity within the next few years.

The human hand is not used just to manipulate objects; it is a critically important sensing device as well. People "feel" their way with unfamiliar objects and situations, and use their dexterous hands to learn about the world around them. It is no accident that children have a near-universal urge to pick up and feel attractive items in a store; touching is one of the most important techniques they have to learn about the new objects they see.

Touch is likely to be an immensely useful sense to an android as well, for it can be used as people do, to help explore the android's world. The sense of touch is also an exquisitely sensitive method of investigating the properties of objects encountered. It can be used to discover a great deal of information about objects, including shape, hardness, texture, and other features; these characteristics are sometimes difficult (or deceptive) to determine from vision alone.

Paolo Dario of the University of Pisa in Italy has developed a robotic finger that shares many characteristics of a human finger. This system is an innovative and original manipulator that illustrates many of the principles involved in constructing a humanlike robotic hand. The robotic finger was developed by first determining what the overall system requirements should be, including decisions about how and what kind of sensory information the finger should detect, and at what level finger motion should be controlled. Let's consider some of these issues for a moment.

Artificial touching systems, unlike most artificial vision systems, must intimately relate control of the motion of the finger with sensory input from the finger. A vision system primarily takes input only, but a robotic finger must correlate the motion of the finger with the sensory data returned from the finger. In effect, motion control of the finger must be processed in two ways: once to generate and monitor the action movements made—a kind of low-level processing—and once again at a higher processing level to determine whatever sensory input can be derived from the motion to enhance the perception of the object.

When a finger actually touches an object, two kinds of sensory information can be fed back to the robot for analysis. One is the global motion of the finger as a whole; this covers, for example, how sharply the forward motion of the finger was stopped by its contact with the object being touched. When compared to the speed at which the finger was moving, this gives some notion of the relative hardness of the object. A second kind of contact information comes from the local sensory input from those finger sensors located in the part of the finger that actually makes contact with the object. The dividing line between these two kinds of inputs can be fuzzy at times. If the robot is to have both these kinds of sensory input, however, it must be provided with a deformable covering. It cannot be a simple metallic gripper as most of today's industrial robots have. It must have a soft, cushioned surface with sensors placed beneath the outer covering that can report on changes in the shape of the finger's surface. There are a couple of ways this can be done: one is to sense the stresses within the surface material itself; the other is to monitor the deformed shape of the outer covering.

For fine motor control (dexterity) of the finger or any other robotic arm, a motion sensor is more accurate when it is closer to the actual part of the robotic system that moves—in this case, the end of the finger. The only problem with this is that this also exposes the sensor itself to accidental damage, and a sensor capable of detecting very fine changes in contact pressure is almost certainly a delicate device. A compromise must be reached in this issue. In the human body, two separate control and sensing systems come into play: A gross-motion system based on the muscles and tendons of the arm, wrist, and hand is used for heavy objects or large-scale motions; delicate movements are controlled primarily by force sensors just under the skin of the hand and fingers. Dario decided that a similar scheme should be used in his robotic finger.

Notice, by the way, that in a finger with the characteristics described here, there is no reason why additional sensors cannot be added underneath the outer covering. One possibility for such supplemental sensors is the addition of temperature sensors. This would enable the android to sense the surface temperature of objects it touches, just as a person does. Another possibility is to add chemical sensors that can detect something of the chemical nature of nearby objects. This might be very much like adding a sense of smell through the fingertips rather than the nasal passages. (Catfish "smell" through olfactory sensors that are spread over nearly their whole body, so such "finger-sniffers" have at least a vague parallel in animal sensory systems.)

Yet another kind of additional sensor might be a pain sensor. This could be used to detect overstresses in any of the other sensors, whether temperature, tactile, or even olfactory. The role such a sensor could play is identical to that in human beings: to prevent damage that might occur if the finger presses a dangerous object, or to limit the damage (if it has already occurred) by prompting the motor control system to remove the finger from the object immediately. A pain sensor could warn the android that the pot it just touched is too hot for safe handling, or that some damage has occurred to the finger because of an unusually sharp edge the finger pressed against.

Once these kinds of system-level issues were considered, Dario's group began building the robotic finger. The sensors used within the finger are constructed from a film of a special material that is both piezoelectric and pyroelectric. Piezoelectric means that the material generates a tiny electrical charge when it is mechanically stressed or deformed. This property makes it a good contact or tactile sensor. Pyroelectric means that the material generates a small electrical charge when it is heated; it thus makes a good temperature sensor. When used in the finger, two separate sets of sensors are used, a surface, epidermal, layer that senses gross attributes of the surface, and a second, dermal layer underneath that is much more sensitive to fine details in the surface.

In addition, the finger has a special sensor on the upper surface that can detect tiny variations in the regularity of the surface; the analogy is like running your fingernail over the striations of a 33 rpm record.

When Dario's robotic finger approaches an object, one of its first actions is to determine what part of the finger actually contacted the object. Fingers are not just two-dimensional objects; they are (more or less) cylindrical, and sensors are placed all around the side of the cylinder. Initial contact can be determined from the array of low-resolution deformation sensors surrounding the robot finger. When the finger touches the object, the outer covering is deformed; it can be assumed that the sensor that experiences the greatest deformation is the one that corresponds to the point of contact. Once that point is determined, it provides some details as to the orientation of the contacting surface, and an estimate of the direction in which to move the finger to slide along the surface. Also, deformations of nearby sensors gives some information about gross surface details of the object.

Dario's robot finger is organized as a collection of modular processors arranged in a hierarchical system that is reminiscent of Kawato's robot arm (though not as complex: it is only a single finger, after all!). Low-level processing is located in or near the finger itself and consists of sensory input devices along with an analog control unit for motion control. In the middle processing level, sensory data is consolidated for later high-level review, and is also provided as feedback to special sub-processors that provide general control for specific kinds of actions. These mid-level actions include functions such as motor control for approaching an object, sensing the shape of a contacted object, determining the texture of the object, and so on. These parallel action-control modules jointly send their instructions to the low-level motor control for implementation. Finally, the highest level of processing integrates the sensory data passed up from the mid-level processor and determines an overall tactile plan. This module can consult a database and do high-level planning for the finger's motion.

Dario's robotic finger looks astonishingly like a human finger. The sensory covering acts as a "skin" and the top-surface striation-detector is designed to look like a human fingernail. While its operation is still being refined, the overall modular, hierarchical approach, combined with a careful consideration of the complete system (rather than just constructing a finger with no concern for sensory and planning integration) gives this system enormous possibilities for an android hand.

Bernard Widrow, a professor of Electrical Engineering at Stanford University, is a charming elf of a man. He has a reputation for turning out first-rate graduate students who go on to make their mark in their fields; in addition, Widrow's lab has turned out some amazingly successful applications over the past thirty or so years. His lab's latest project, however, may prove to be the most stellar of all. Working in conjunction with Dr. Joseph Rosen of the Stanford University Division of Surgery, two of Widrow's graduate students have begun a breakthrough project. Eric Wan, a PhD candidate in electrical engineering and a medical student at the School of Medicine, has worked on developing a neural interface chip. With the assistance of Gregory Kovacs, another PhD candidate in electrical engineering, and under the guidance of Drs. Rosen and Widrow, the two have developed a silicon neural network chip with tiny "holes" along the edge for connectors.

The idea behind the chip is this: When a person suffers severe damage to an arm, perhaps including partial severing of the limb, surgical repair can often reconnect the main nerve fibers. However, during recuperation individual nerve cells experience a scrambling of their signals, with the result that full use of the arm is never completely regained. When the neural interface chip is implanted, however, the nerve fibers grow through the special connector "holes" along the edges of the chip and make an electrical connection to it. These electrical connections—to the shoulder nerves leading to the brain on one side, and the hand and arm nerves on the other—act as input and output signals to a small artificial neural network embedded in the chip. Actual experiments with rats and monkeys (using nonfunctional chips) have confirmed that this growth and connection process does occur. Furthermore, the researchers have confirmed the establishment of two-way communications with the nervous system, as well as the fact that the chip does not degrade in its biological environment, even over long periods of implantation. Once this chip-nerve interface is complete, training can begin.

The chip's on-board neural network has the function of rerouting the scrambled signals so that the patient can regain control of the damaged limb. In effect, the network is to act as a translator between the nerve fibers, so that the hand and arm coordination the patient knew before the accident can be regained. This process has been carried out in simulation, with the result that with less than a thousand training trials—each trial consisting of a single command to the hand or arm such as "clench the fist"—all nerve communications were completely unscrambled, implying the return of 100 percent of the hand's functionality to the patient.

The possibilities are even more interesting when we consider the construction of limb prostheses. Current artificial hands are awkward and hard to use. The latest technology permits the patient to use shoulder muscle movements to control the hand; however, this requires constant concentration on the part of the patient, and thus the prostheses are fairly unreliable—too much perspiration can cause sensor failure and thus make the arm difficult to manipulate, for example. If the embedded neural network chip could be combined with a high-quality mechanical prosthesis, then the control of the artificial limb would appear to the patient to be as natural as moving his or her own hand and arm. Since exactly the same nerves would convey the arm control instructions as before—as well as the sensory data coming back from the arm to the brain—the patient should be able to use the prosthesis with completely natural coordination and control.

Does all this sound far-fetched? Perhaps, particularly if you have seen the second *Star Wars* movie (*The Empire Strikes Back*), at the end of which Luke Skywalker was fitted with just such a prosthesis for his severed hand. But don't bet against it, because Widrow's laboratory expects to be able to create fully functioning prostheses before the turn of the century.

We have strayed just a bit from the discussion of arm and hand control for an android, but the technology that manufactures neural-implant prostheses is likely to also be able to help in manufacturing coordination and grasping systems for an android. And when combined with the innovation of Kawato's feedback control robot arm, Dario's human-like robotic finger, and Neurogen's INFANT system, as well as other current and future developments in the field, it seems reasonably certain that we can expect immensely sophisticated robot-arm entities in the next five to ten years.

So perhaps then we will be able to deal with a more philosophical version of the question posed by Browning: Should an *android*'s reach exceed its grasp?

≧ 5 ≦

Remembering the Past . . .

Computers are useless.
They can only give you answers.
Pablo Picasso

Mark Twain once confessed that "my memory is loaded with nothing but blank cartridges." Many people occasionally (or frequently) feel that way. Yet human memory is an often-maligned system that performs astonishing feats on a daily basis. We can recall people, places, and events seen only once; we can memorize the sound of a popular tune, the feel of velvet, the taste of chocolate, and the scent of an enticing perfume. And we do all this automatically, with little deliberate effort, and with amazing accuracy. While the police may moan over the unreliability of eyewitness accounts to a crime, the truly astonishing fact is that detailed memories are retained at all. Given the complexity of stimuli that constantly bombard us every second, the human memory does a wonderful job of storing relevant information.

How does all this work? Human memory is not fully understood, but the general outlines of how it operates are known. There are at least three different kinds of memory systems in the brain: sensory memory, short-term memory, and long-term memory. A common belief among psychologists is that these three memory systems pass information among each other through some special processing systems such as an attentional system, and one or two kinds of rehearsal systems. Let's see how all this works together.

Sensory memory stores the sensory information that bombards us constantly. It preserves the "raw data" of our information processing system, and because it does so, it must have a very high storage capacity overall. Why is this so? Consider for a moment a visual image. Suppose we imagine a typical high-resolution video monitor. The entire image on the monitor can be perceived at once by the eye (although individual areas may have to be studied separately for full comprehension of the image details), so visual memory must retain at least that much information. A picture consisting of, say 1024 image dots (called picture elements or pixels) across the screen and 1024 pixels down has a total of 1024 × 1024, or approximately one million, total dots in the image. In a black and white image, each of these picture elements could be either black or white, resulting in the need to store a 1 or a 0 for each of these dots, 1 if the dot is black, and 0 if the dot is white. Thus the total memory needed to support the video monitor is about one million binary (2-valued) digits. But suppose the monitor is not just black-and-white, but permits gray-scale images. In this case even more information must be stored, since every dot might have any of a number of levels of darkness, or gray levels, from 1 (black) to 0 (white). The video monitor certainly has some limit to its memory, and this is usually specified by the number of gray levels it can support at each pixel location. If it permits 16 levels for each pixel, it takes 4 binary digits (bits) to define the shade of gray at each location (2 to the fourth power is 16) and thus the storage needed to support it is 4 × 1024 × 1024, or about four million bits. The human eye can distinguish many more shades than this, however, so for a realistic picture we might double the memory needed and provide 8 bits for each pixel location. This increases the number of grays available at each pixel to 256 (2 to the eighth power), and increases the storage required to 8 × 1024 × 1024, or about 8 million bits. Now suppose we want to make the image in living color. Generally color monitors operate using separate memories for each of three "guns," one for red, one for blue, and one for green, the three primary colors of light. A pixel's color is determined by the combined intensities of each of the three guns at its location, and literally any color can be generated by various combinations of these three basic hues. Thus a choice of 256 color levels for each of the three guns at each pixel location generates 256 × 256 × 256 possible colors, or a total of about 16 million possibilities! It also requires not 8 million bits of memory, as did the gray-scale monitor, but three times that amount to store the pixel intensities for each of the three guns.

While these numbers are very large, they only serve to drive home the point that the raw image data handled by sensory memory must have a very high capacity indeed. The human eye is capable of processing even finer images than those of the video monitor described. A person with good eyesight can process an image with about 10,000 × 10,000 pixels. This means that at the human eye's resolution, an image needs about *2.4 billion* (2.4×10^8) bits, or about 300,000,000 bytes, of memory to be stored completely. (One byte equals 8 bits of memory.) To put this into perspective, one of today's high-end, top-of-the-line workstations has perhaps 8,000,000 to 16,000,000 bytes of total memory capacity, or only about 2- to 4-percent of the memory needed to store just a single image.

Furthermore, there are other senses as well, all of which are continually transmitting raw data to the brain. Sound is another sense that demands extraordinarily large storage capacities, and the senses of smell, taste, and touch also demand their own large storage capabilities.

Sensory memory, then, must be able to handle very large amounts of information that arrive almost continually. But because the storage requirements are so huge, it cannot retain the information it stores very long before it must "write over" it with newly arrived data. Current estimates are that most sensory information in only saved about a second before it is forgotten, except for the sense of hearing, which seems to retain information for about three seconds. When you are suddenly "brought to attention" in a boring conversation and mentally "playback" the last remark so you can respond to it, you are experiencing your sensory memory in action.

Short-term memory, generally abbreviated STM, provides a sort of erasable tablet for brief storage of information. A common example of STM occurs when you look up a phone number and do not write it down or deliberately memorize it. Typically, it can be retained only for about 15 or 20 seconds before fading away. Also, STM doesn't appear to hold much information. Through list-learning experiments where participants are asked to remember lists of unrelated words, and many other tests, it appears that we can store from 7 to 9 items at a time in STM. This is the reason it is occasionally difficult to recall a phone number, even immediately after looking it up—the 7 digits of a telephone number are right at the boundaries of our STM capacity. Very few people can correctly recall the 10 digits of a long-distance number (assuming an unfamiliar area code) without checking it at least once while dialing it. Similarly, 9-digit zip codes continue to meet a great deal of resistance from the general public for everyday addresses—they are definitely at or beyond the edge of human STM capabilities. This is also why it is easier to remember long strings of digits as sequences of 2- or 3-digit numbers rather than recalling each digit separately. For example, it is easier for me to recall my house number as eleven-four-fifty than it is to recall it as one-one-four-five-oh. In the first case I store only three numbers; in the second, five.

The third kind of human memory is long-term memory, or LTM. This is what our memories are truly made of, and it is the system most people immediately think of when asked about "memory." LTM is that which remembers who and what we are, what we are doing, and where we come from. It is truly what makes us human. It has a vast capacity, because the memory of a healthy person never "fills up," even with a lifetime of experiences to store. Unlike sensory memory and STM, however, information is not stored in long-term memory immediately. Instead, it must be filtered by the other two memory systems before entering such permanent storage.

One widely held current human memory model indicates that there are at least two systems that monitor the exchange of information among the various memory systems. A system referred to as the "attentional" system controls the passage of information from the sensory memory into short-term memory. The attentional system explains the phenomenon of being able to recall data better when you're paying attention than when you're not. A second system, called the "rehearsal" system, controls the information flow between short-term and long-term memories. This flow of information is bidirectional—recalled information flows from long-term memory to short-term memory for current use, and information to be stored flows from short-term memory to long-term memory for permanent storage. The rehearsal system is called that because it models the behavior of thinking about or rehearsing an action or event either before or after it occurs.

For the moment, consider only long-term memory, since it is the one that permanently stores information. No memory system worth its salt can operate unless it performs at least three functions. It must encode the information presented for storage in some reasonable fashion that can be processed (or this step must be performed externally before the information is presented to the memory). It must store the information. It must recall the information on demand. Just as the human brain must have these abilities, so must an android. It is time now to consider how we might build an artificial system with the memory capabilities we need.

The obvious model to work from is computer memory. As was noted in the first chapter, a digital computer's memory is much like a collection of pigeonholes. Each memory location has a specific label, called its address, and each can contain exactly the same amount of information. Since these are digital computers, one might assume that the information would be in the form of digits, and this is of course correct. A computer must have information it is to remember coded as a sequence of binary digits. Essentially, a computer really only knows two concepts: 1 and 0. Everything it knows must be in codes made up of these two digits.

Generally, a computer memory location stores exactly one word of information, where a word is a collection of bits of a specific size. Until recently, the most popular word length has been 16 bits; today that has increased to 32 bits, as found in most advanced personal computer systems. Supercomputers may have word lengths of up to

64 bits or more. To store information that is non-numerical, it has to be coded as numbers so that the memory can handle it. Thus, the letter *A* is stored as "0100 0001" when coded in the commonly used ASCII code. All instructions and data stored in the computer must eventually be encoded as 1s and 0s in a similar fashion.

In order for the computer memory to be effective, there has to be some means of deciding in which pigeonhole to place a particular word of memory. Furthermore, if we ever want to be able to retrieve that word, we have to know exactly where it is so we can get it back out. Several techniques have been developed to do this. By far the most common simply places the information in memory in a sequential fashion. In other words, if we know that the desired word is the 354th word in a program, and if the program uses sequential storage beginning with the 165,468th word of memory, then we can easily retrieve the value by looking in memory position 165,822. This is an extremely simple method of memory storage, and has been used since the very first digital computer. It works quite well, even if it is not a particularly intelligent system. For example, this system could (and generally does) store completely unrelated items next to each other. In fact, the final position of a value within the memory is more or less random (although well defined once it is stored there). There is no particular rhyme or reason to this system at all.

A second common technique is through the use of pointers. These are special memory locations that do nothing but keep track of where a particular value is stored. For example, suppose the value of *A* resides in memory cell 384. Because sometime in the future *A* might move somewhere else in memory (because of some kind of processing the program does or because of the needs of the computer's operating system), a special memory cell called *A** is also set up. *A** is in a known, fixed location that never changes, and the value placed in it is 384. Thus two ways exist to find *A:* looking directly in cell 384, or looking at *A** and using the value there to determine *A*'s location. Obviously, using *A** is a two-step process (usually referred to as an "indirect" reference to *A*) and furthermore it means that *A* now needs two cells to keep track of it instead of just one. But if the program being executed causes the location of *A* to move around (as often happens in many computers and computer languages), then having *A** available as a more permanent access route is an excellent decision in spite of the additional time and space overhead that results. When *A* moves to a new memory location, all the computer has to do is to change the value in *A** so that it reflects *A*'s new position im memory. With this system, no matter how many times *A* moves, or where it moves to, we can always find it just by looking in *A** for the current position.

This clever pointer system is somewhat more recent than the simple sequential system above, but it still is based on the same premise. At least initially, we have to know exactly where everything is in memory. Pointers can be complex to program with, and they do nothing to help organize memory by storing like concepts together. In fact, using pointers just allows memory to be more disorganized than ever as values jump randomly around the possible memory locations.

Databases are a special kind of computer program that provide the ability to store related pieces of information together in a more meaningful manner. There are all sorts of database storage techniques—they are the most studied kinds of computer programs in the world—but the one of interest here is a system called hashing.

Databases allow the creation of a format for information entry and then store all related items in a simple grouping called a record. For example, a business might use a database to keep track of its customers, and so would want to store all the information about that customer in a single place. Thus, whenever customer "James Bond" is recalled, a record of his preferences for fast automobiles, dangerous women, and elegantly tailored clothes would be recalled; "Indiana Jones" might recall preferences for bullwhips, a battered fedora, and an ancient leather jacket. And "Batman" might recall unusual custom cars, a black cape and mask, and an odd penchant for running around in longjohns. Unlike the simpler computer memory systems, all these related pieces of information about each person are stored together, so that they can always be recalled together. But because we don't know ahead of time just how many different people we eventually have to keep records on, the memory organization scheme has to be more sophisticated than anything described so far. If they are just placed sequentially in the (more or less random) order in which they are received, it is difficult to remember, for example, whether Batman was the second or third entry. While a table of pointers might help this a little, if the database contains thousands or tens of thousands of entries, searching through that table to find a particular person takes longer and longer to do.

Suppose we want to add "Han Solo" to the list, storing preferences for hotrod spaceships, money-making schemes, and rebel princesses along with it. Where should this record be stored, given perhaps 500,000 possible memory locations to choose from? Using a scheme called hashing, all (or certain key parts) of the record are encoded as a number. This number is then used as input to a special function called a hashing function, the output of which is used as the address of the record for Han. The hashing function is nothing more than a special mathematical operation that converts its encoded input into a single number that can be used as an address. It (along with the technique used to encode the record into an input for the hashing function) also has the property that each unique record is likely to generate a unique address, so if we later add "Thomas Magnum" who prefers fast cars, dangerous women, and comfortably sloppy clothes, it is not likely that the hashing function generates a duplicate entry, even though James Bond shares several of the same preferences.

Of course, occasionally a hashing function causes a "hit" by specifying that a new record be added in an already used location. Obviously the system must have a check for this situation so that the old record is not written over by the new one. However, if the hit rate is sufficiently low, the addition of a few well-phrased rules allows the system to recover nicely from these exceptional circumstances. And the advantage gained is one of access speed. Even with thousands or tens of thousands of entries, any given entry can be located by encoding its key elements, and passing the result through the hashing function. This is really just a more sophisticated version of the simple pointer system, but it is so effective that many database systems have used this technique.

Unfortunately, none of these systems really allows information to be stored in the same way people store memories. The android needs a system that puts related concepts together in actuality, not just as a superimposed structure on what is basically a sequential physical memory. The android needs an associative memory system.

Associative memories store information by keeping it physically next to similar information. Many human memories seem to be stored this way, and the result is a highly accessible, flexible storage system that works (most of the time!) marvelously well. People often associate items they want to remember with other, highly memorable items—this is one of the key techniques taught in memory improvement courses. For example, suppose you want to remember a grocery list. One way is to simply try to memorize the list; this may or may not work well, depending on your skill in memorizing, your motivation, and the length of the list. A more successful technique—assuming you cannot just write down the list on paper!—is to associate something with every item on the list. A number and rhyme scheme could be used like this: "The first is juice to quench my thirst; two is cat food for you-know-who; three are cookies all for me; four is milk for me to pour," and so on. Making up silly little ditties like this makes it easier to remember each item. Another associative memory technique is to associate each item with either a tour of a house (or grocery store) or with something exceedingly silly. An example, using the same brief list as before, might be to visualize a calf opening the refrigerator for milk, while playing kickball with cans of cat food and frozen juice, and trying to wipe cookie crumbs off its face with its tongue. It doesn't matter what the image is that is used as an association, as long as it is silly, funny, or otherwise easy to recall.

People use associative memory techniques even when they are not formally trained to do so. For example, it is likely that if you were over the age of 8 or 10 at the time, you can instantly recall exactly where you were when you heard that John Kennedy had been shot, or what you were doing when you first heard of the attack on Pearl Harbor. Even though, for most people, these two days were relatively normal days, because they are associated with emotionally charged and memorable events, the days stand out in our minds. On a more mundane level, if a young woman is asked who her closest boyfriend was in the summer of 1983, she would almost certainly figure it out by thinking something on the lines of: "Let's see now . . . in 1983 I was just starting in my senior year in college, and that summer I worked at the carnival, where I met that cute redhead who lisped and followed me around for days. Now what was his name?"

We associate people and events we know with the places or situations where we are likely to encounter them. This is why we occasionally have trouble recognizing people when they are encountered under unexpected and unusual circumstances. If we don't associate the person with that location or circumstance, we have a harder time recalling his or her identity. While occasionally embarrassing, it is a perfectly natural result of our associative memory storage technique.

Associative memories have some characteristics that people find highly useful. One of these characteristics is called "robustness." Actually, associative memories are robust in two distinctly different ways. The first is that they can correctly recall information based on incomplete or garbled cues. For example, if someone is asked to recall the name of a world leader with an irregular birthmark on his forehead, few would have trouble identifying him as Mikhail Gorbachev. Yet the amont of information provided—male, world leader, with a birthmark—is quite minimal. In spite of the incompleteness of the description, recall can be perfectly performed. Even when the recollection cues are garbled or slightly inaccurate, the memory can still be retrieved; this helps us to recognize a friend who has recently grown a beard or changed to a new hairstyle, or to understand the words of someone with a bad cold or a strong accent. We correctly interpret even such garbled inputs and make the proper identification.

The second way associative memories demonstrate robustness is in their tolerance for what might be termed "hardware failure." Associative memories generally do not work like the traditional computer memory in which each fact is kept in its own identifiable pigeonhole. While this makes a rather neat and tidy memory system, if something happens to that pigeonhole (or if, for any reason, the system loses its ability to find the correct pigeonhole, as can happen with a stray pointer), then the fact stored in it is effectively erased or forgotten. The pigeonhole memory system suffices for a computer because it is frequently monitored for any hardware memory problems. It is not at all unusual for a computer to do at least a cursory check of its memory every time it powers up; some systems run complete diagnostic programs on memory regularly. Still, loss of one or more memory locations is one of the more frequent causes for computer breakdowns.

The problem with putting all the storage of a given fact in one location in a biological system is that this system is far too fragile, especially given the fact that hundreds or thousands of neurons die daily. There is an old, and now discredited, theory of memory that assumed that one neuron in the brain had the responsibility of remembering each fact learned, as is the case in a computer. The classic refutation of this theory is that since thousands of neurons die every day, if this system were in use you would one day wake up and not be able to recognize your own grandmother because the cell storing her image had died. Since this does not happen, such "grandmother cell" storage cannot be the case in the brain. And, in fact, it is not.

Associative memories nearly always store information in a highly distributed fashion rather than in the grandmother cell style. This means that the memory for a given fact is spread over many locations in the memory; thus, if one memory location fails (the neuron dies), then the fact is not lost at all, since it also exists in other locations as well. The perfect example of this kind of distributed storage is a thesaurus. Suppose you take a black marking pen and strike out the entry for the word "commence" in a thesaurus. Is that word completely lost? No. If you look up the word "begin" and check its list of synonyms, you find "commence" stored there as well. Similarly, if you look at "start," or "inaugurate," or any of a number of other words and phrases, you find "commence" listed in each place. The total storage for the word "commence" is distributed throughout the thesaurus, and is physically located next to those other entries that are in some way associated with it. The thesaurus is thus an excellent example of a distributed associative memory.

One implication of such a distributed storage system is that an associative memory tends to be somewhat insensitive to minor failures in memory components. Because no single storage location bears the entire responsibility of remembering anything, if one or a few locations fail, little is lost. This leads to another nice characteristic of associative memories: graceful degradation in performance. This means that as the memory experiences multiple or repeated hardware failures, its performance gradually declines from peak levels. Rather than experiencing a sudden, complete loss of function, an associative memory, like an old soldier, just slowly fades away, being able to recall less and less over a long period of gradual decline. Without this characteristic, no biological system would be able to function much after early childhood.

As an interesting side note, it is useful to consider just how much loss of function is involved in the brain over a normal human lifetime. The typical human brain has about a hundred billion (100,000,000,000) neurons at birth. Each person loses a few thousand of these every day, and over a lifetime of, say, a hundred years, loses perhaps two billion (2,000,000,000), assuming an average 5000-neuron loss each day. (The use of even small amounts of drugs or alcohol, or the presence of a number of diseases, can drastically increase that daily loss factor, but I'll assume a healthy teetotaler here.) That involves a total "hardware" loss of about 2 percent. Nearly any artificial neural network can sustain that level of failure with little or no perceptible loss of performance. Furthermore, while the neurons in the brain's biological neural network generally cannot reproduce, they can grow new connections to other neurons, so even the 2 percent hardware loss can be compensated for by changing the connections between the remaining neurons. This is one of the circumstances that occurs when a stroke victim is retrained to speak or walk, for example; sections of the brain that are undamaged are trained to take over the lost functions of the damaged sections. As a consequence, teenagers cannot use this argument to explain why they are clearly so much smarter than their parents!

Associative memories have several characteristics that are highly desirable: they are robust in the sense of correctly processing garbled and incomplete data; they are robust also in the sense of gracefully handling reasonable amounts of hardware failure; they store memories distributively like a thesaurus. They also have one more characteristic that is extremely useful: they are generally content-addressable.

Content addressability is a computer term referring to a memory that determines the storage location of a fact by the contents of the fact itself. The hashing scheme outlined above is one simple way a computer can implement content addressability. More generally, though, content addressability means that a particular fact is accessed by presenting a (usually incomplete or garbled) version of the fact. So when you see Mickey Mouse's ears, you don't have to see the rest of his face to recognize him. Or when you hear Miss Piggy say "moi" in her enchantingly shrill voice, you don't need to see her to recognize who she is. As noted before, this is one of the characteristics of human memory, and thus one that should be incorporated in any android memory.

All these general characteristics are very nice to have, but is there any way to actually build one? The answer to that is a resounding yes. Researchers have been developing associative memory systems for years, and some of them are quite good.

One simple way that associative memories can be constructed is to build something that resembles a telephone switching system. Key versions of these systems were developed by John Hopfield of California Institute of Technology and later extended by Bart Kosko of the University of Southern California, among many other researchers. In essence, the associative memory systems consist of nothing more than a collection of nodes in which every node is connected to every other node. (The analogy comes from the switches in the telephone system, in which every phone can contact every other phone in the system.) These "crossbar" systems, named after the telephone exchange switches, can be built to associate collections of pattern pairs.* Each pattern pair member is associated with the other member of the pair, so that in the pair (*A*, *B*), *A* recalls *B* and *B* recalls *A*.

Crossbar systems, and their slightly more complex cousins bidirectional associative memories, are one of the very simplest kinds of artificial neural networks because they generally don't have an iterative learning law associated with them. Essentially they consist of a one- or two-dimensional matrix of neurodes in which every neurode connects to every other neurode. The patterns that are to be stored determine the strength of the connection paths between the nodes in the network, so that, in essence, the memory is built already knowing the patterns it is to store.

Since there is generally no ongoing learning involved in a crossbar network, these are also the easiest associative memory systems to build. Each pattern to be stored is encoded as a vector, a column of numbers that represent the various features of the pattern. To associate two patterns together, the two corresponding vectors are multiplied together in a special way that results not in a single number, but in a two-dimensional array of numbers. These numbers represent the connection strengths that store that pattern pair. To build the entire memory storing a collection of pattern pairs, the two-dimensional arrays from each pattern pair are simply added together, giving the final connection weight, or storage, matrix.

A memory system isn't of much use unless it can recall informa-

*Another name for single-layer crossbars is "Hopfield networks," named after John Hopfield; bidirectonal associative memories are frequently abbreviated to "BAMs."

tion as well as store it.* It turns out that recall operations are done almost as easily as building the memory. Given one-half of a pattern pair, pattern *B* perhaps, to recall the other half, *A*, you just multiply *B*'s pattern vector by the storage matrix. The result is a vector of the proper size for *A*, but it may not necessarily be the final answer. To confirm that the answer is correct, the storage matrix is multiplied by this vector to see if the result is *B*. If so, then the vector generated in the first step should be *A*. Otherwise, this new *B*-sized vector is used again as input for another round of matrix-vector multiplications. The concept is one in which the multiplication results stabilize so that a single consistent vector pair no longer changes; once that happens, the result should be the pattern that was stored in association with *B*. Generally speaking, these memories correctly recall *A* after only a few multiplication passes.

For those who are not familiar with matrix mathematics, this may seem to be a bit like magic, but it works reasonably well for such a simple system. Nevertheless, problems exist with such crossbar memories. The first is that they are not very efficient in storage space. To store pairs of numbers, each number in the pair is coded as binary vector patterns. For any number up to 1000, ten binary digits are needed to encode the number. (Two to the tenth power is 1024, so ten binary digits can encode up to 1024 distinct numbers ranging from zero to 1023.) That means that the pattern vectors have 10 elements apiece, and when multiplied together to make the storage matrix, there are 10×10, or 100, elements in that matrix. If built as a computer simulation, it requires 100 memory pigeionholes just to remember the storage matrix, using one address for each connection weight. If built directly in hardware, it requires 100 wires (or circuit traces) to accomplish the same effect. Suppose 10 pairs of numbers are stored in the memory; in effect, the memory then holds a total of 20 numbers. But 100 memory locations have then been used to hold those 20 numbers! Granted, the numbers have truly been stored associatively, with each one associated with its counterpart, but this is not a particularly efficient storage medium nonetheless.

If it is so inefficient, why not just store more pairs of numbers in the associative memory? Unfortunately, this does not work either. It turns out that the more pairs of numbers that are stored in a crossbar, the more likely it is that they interfere with each other. For example,

*One of the classic pranks that computer science students like to play on naive computer users is to tell them about the development of a new WOM chip that implements an innovative Write-Only Memory scheme. Such a "you can put it in but you can't get it back out" storage system is unlikely to be commercially available in the near future—at least not intentionally so.

the pattern (1 0 1 0 1) would not interfere at all with the pattern (0 1 0 1 0). But it would likely cause severe problems if the pattern (1 1 1 0 1) were stored in the memory as well. There is simply too much overlap between (1 0 1 0 1) and (1 1 1 0 1) to permit correct recall. As a result, if both are stored in the same memory, one (or both) of them is apt to generate a nonsensical recall instead of their associated patterns.

In a sense, this is very much like human memory failures. Psychologists are now becoming more convinced that most, if not all, "losses" from LTM are a result of new memories interfering with older memories (or vice versa), rather than having older memories somehow weaken and decay. Unfortunately, this particular human characteristic is one we would like to avoid in an artificial system.

Crossbar networks can hold no more pattern pairs than the size of the vectors that make up the patterns. So in the example above that used 10-number vectors in coding the patterns, the resulting crossbar memory can store no more than 10 pattern pairs. Worse, this is the *theoretical* storage limit; the actual limit (for patterns that are not carefully chosen to be non-overlapping) is something like 10 to 15 percent of this, and gets smaller as the vector size gets larger. And worst of all, each storage or recall operation involves multiplying matrices and vectors, which, in a serial digital computer, becomes more time consuming according to the square of the size of the input pattern. So increasing the size of the input pattern vector to overcome the storage limitation causes the network to be correspondingly slower unless the memory is implemented directly in parallel hardware. Clearly, we need to either have enormous pattern vectors and simultaneously get around the speed problem, or we need to use another method of storage.

In fact, either of these may be the answer. Researchers at a number of universities and institutions are hard at work trying to develop optical associative memories. These systems use special materials, call spatial light modulators, or SLMs, that vary their optical properties (such as their transparency) according to the electrical voltage across them. Such materials can act as a filter to tiny, microchip-sized lasers. One simple version of this idea, illustrated in Figure 5.1, is to set up a column of tiny lasers of varying intensity, using the laser beam's intensity to encode the value of each number in the input-vector pattern, and pass the laser light through relatively simple cylindrical lenses. The lenses spread each laser's beam into a broadened row of light that then passes through the SLM. Each unit area of the SLM has its optical properties set so that it represents one element in the storage matrix. The laser light is modified by its passage through the SLM, and collected by another cylindrical lens on the other side. The result

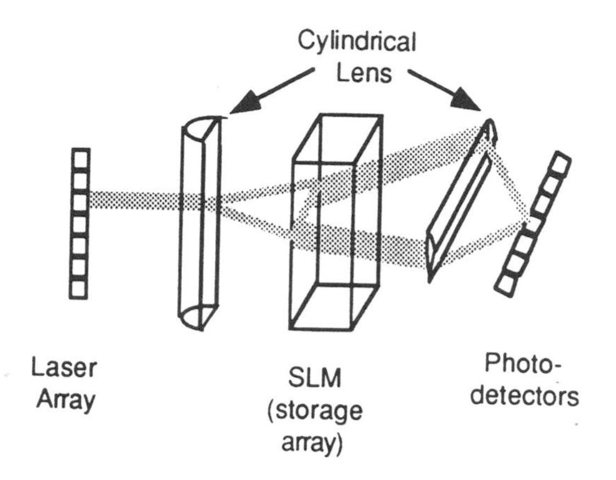

Figure 5.1 An optical implementation of a crossbar associative memory.

is a row of light intensities that corresponds to the pattern vector associated with the original input pattern.

One advantage an optical storage system has over more traditional technologies is sheer volume. Just as today's compact disc technology offers storage densities that are orders of magnitude greater than magnetic media disks, holographic memories promise to be able to handle huge amounts of data. If we can handle large enough matrix arrays, we can overcome the crossbar network's low-density storage problem. Furthermore, as the example above demonstrated, memory recall operations occur literally at the speed of light, no matter how large or complex the pattern vectors are—it only needs the time it takes for light to travel through the system to recollect the associated pattern. (While the speed of light is slower in lenses and SLMs than it is in a vacuum, it is nonetheless extremely fast.)

There is another way of overcoming the problem with crossbar networks, however, and that is to use a different kind of associative memory. Generally speaking, most artificial neural networks can be considered an associative memory because neural networks associate an input pattern with an output pattern. Crossbar networks, while the simplest neural network to build, are also the most obvious. After all, if we contend that the power of the network is in its connections, then what could be more powerful than having every neurode connected to every other neurode? As it turns out, the answer to this is "almost anything."

Crossbars represent a sort of "brute force" approach to neural networks that is likely to be unfruitful in the long run. A better source of inspiration is not this egalitarian, everything-connects-to-everything system, but rather the human brain. Study of the brain has amply demonstrated that it is not a homogeneous, fully connected system; it is made of a large number of highly structured modules, each of which has a particular function, and each of which is specialized to do its job well. This specialization arose as part of the evolutionary history of our species; those brain functions that arose earliest are the most efficient—functions like control of breathing and other autonomic systems fall into this category. Those functions that are more recent in origin, like speech and language and some aspects of control of the hand, are less efficient and take more "brain power" for us to accomplish.

Crossbars have another fatal flaw that is corrected in most other neural networks. Crossbars have at most two layers of neurodes that interact with each other. In other words, an input pattern stimulates the input layer, which is directly connected to the output layer of the network. This means that the kinds of patterns the network can learn are inherently limited. A crossbar network, or any other neural network that has only one or two layers, cannot learn all kinds of associations. They are limited to those problems where the input patterns come in distinctly separate categories (in other words, problems that are linearly separable). If there is a fuzzy or overlapping boundary between categories of input patterns, the crossbar and its cousins fail.

This flaw can be easily corrected, however, by the simple method of adding a third, or middle, layer to the network. Thus the input pattern connects to this middle layer, which in turn communicates to the output layer. When a neural network is arranged in this fashion, there is little or no restriction to the problems that it can solve. In fact, there is even a theorem that proves that a three-layer network exists that can solve any such association problem at all. Unfortunately this theorem, called Kolmogorov's Theorem after the Soviet mathematician who first proved it, does not exactly tell us how to find the network to solve a given problem; as a result we have all sorts of learning laws that slowly zero in on a suitably competent network. It does seem clear, however, that neural network associative memories are the way to go.

While the technology to build large-scale neural networks in hardware is in its infancy at the moment, commercial neural network chips are now available, and more appear on the market every few months. With their true parallelism, they embody a dramatic escalation in our ability to construct compact mobile androids. As an example of the relative processing enhancement such chip products provide, one manufacturer* claims that their chip can learn in only 6

*Adaptive Solutions, located in Beaverton, Oregon makes this claim for their chip, based on relative training times for a standardized version of the NetTalk speech generation problem. NetTalk is described in detail in Chapter 10.

seconds what takes a Sun workstation-based simulator four-and-a-half hours.

Neural networks are already moving into the mainstream of commercial and industrial applications. They are now in products ranging from automobiles to horse-race handicapping systems to airline baggage machines that check for plastic explosives concealed in luggage. With the addition of truly parallel chip implementations, network-based associative memories will almost certainly keep future androids from mimicking Mark Twain's complaint. There will be just no excuse for an android to claim that its memory has "gone blank."

≧ 6 ≦

. . . As a Lesson for the Future

The human mind treats a new idea
the way the body treats a strange protein;
it rejects it.

P. B. Medawar

Long ago in a land far away, a certain peasant achieved remarkable success in training animals. Unfortunately for him, this particular peasant also had a remarkable mouth, and word of his abilities soon traveled throughout the land. The news eventually reached the ears of the king, who called the man to his presence.

"Are you really as wonderful an animal trainer as they say you are?" the king asked.

Not being overly endowed with either common sense or modesty, the commoner replied, "Oh yes, Sire, I can train any animal to do anything I like."

The king was not a fool, nor did he suffer braggarts lightly, so he said, "In that case, I want you to train my favorite horse to sing an aria. And if you fail I will have you beheaded."

The peasant was a quick thinker, and assured the king that in spite of the difficulty of the problem his wishes would be carried out, and within a year and a day the horse would sing. But when he went to his home to pack his belongings and told his friends about his new task, they were aghast. How could he be so foolish as to believe that he could train a horse to sing an aria?

The peasant was confident as he replied, "I have a year and a day to carry out my task. Many things might happen in that time. The horse might die. The king might die. I might die. Or the horse might sing."

The optimistic peasant in this story recognized the need to be able to train animals to do what he wanted. Similarly, an android, even one that remembers events, doesn't do us much good unless we can make it do what we want. There are two possible ways that it can be made to act the way we want it to. We can program it, or we can train it. Consider the differences between the two.

Computers are made to perform a task through the process of writing a computer program. This is a sequence of very specific instructions (and the data they require) that detail exactly what the computer is to do, and specify the exact order in which they are to be carried out. The task of writing the program can be accomplished in many ways. The traditional approach is a multistep process of specifying the requirements for the final system through consultation with the final user, designing the system, translating the design into a computer language, testing the resulting program, and correcting any errors found in the process. This process is time consuming and prone to errors, particularly if the program is complex.

The other difficulty with writing a traditional computer program is that computers tend to be inflexible. The computer must be told exactly what to do under all circumstances, which means that the programmer must consider in advance what those circumstances might be, and provide specific instructions for each case. Generally speaking, the more flexible the computer's behavior has to be, the more difficult it is to write a program to cause that behavior.

It would be very nice if we could build a system that is capable of being trained to do a task. Training implies that the system can generalize from specific examples to more general principles. For example, if we train a puppy to heel, we reward examples of behavior that are acceptable and withhold rewards for behaviors that are not. From specific examples of acceptable and unacceptable actions, the puppy eventually learns how to heel. This of course demands that the puppy be given feedback that indicates whether a particular action is acceptable or unacceptable to us. These same principles of training should hold with an android. It should be able to learn from examples (and the feedback from those examples) what to do.

Training and learning are not really the same, though they are often confused. Training is an external process that results in learning; learning is the series of internal changes, often caused by an external training program, that are detected by behavioral or other observable changes. Training is the deliberate program of events that the animal experiences and that results in learning, which is detected by changes in the animal's behavior. Training is usually a repetitious process. Just as schoolchildren are drilled on the multiplication tables, training usually involves practice and multiple presentations of the lesson to make sure that it is properly learned.

Three basic kinds of training can be used: supervised, graded, and unsupervised. Supervised training provides the student with exact feedback on the results of a particular training example. It occurs when a student reads aloud, and has his pronunciation corrected while he reads, for example. The teacher explicitly tells the student exactly how well or how poorly each word is pronounced; only if the word is said correctly is it not corrected by the instructor. Graded training occurs when the student is provided with feedback, but it is not exact or explicit. For example, a child playing a simple number guessing game is told her guess is "too high" or "too low" and from these clues must deduce what the correct number is. (If supervised training is used in the game, the child would be told she is "five too low" or "too high by seven"; this would of course take all the fun out of the game.) Unsupervised training occurs when there is no instructor to provide feedback; it is generally associated with a special kind of learning called observational learning. The child sees something happen, or does something by accident, and on her own, decides to learn how to do that action again. This is very sophisticated behavior and we consider it carefully later.

We can build trainable artificial systems; neural networks nearly always have this characteristic. For example, the most commonly used network of all is the backpropagation network, originally described in the 1970s and brought to prominence in the 1980s by James McClelland, David Rumelhart, Geoffrey Hinton, and many others. It uses a supervised training procedure to learn to associate patterns. Since backpropagation is the classic neural network example of a system that learns during supervised training, it is useful to consider its operation in detail.

Suppose we want to train a backpropagation network to perform a task—perhaps recognize the letters of the alphabet. We begin with a network that has three layers: an input layer, an output layer, and a middle or hidden layer, as shown in Figure 6.1. The input layer is presented with an image of a letter and we expect the output layer to produce an identification of the letter. Suppose, to be specific, that the input layer is a 5×7 grid of dots that display an image, and the output layer has 8 elements, enough to display the binary ASCII code for each letter. So if the network believes the image on the input grid is a picture of the letter *C*, the output elements are supposed to reproduce the ASCII code for *C*, 01000011. In the middle layer, the network has 12 neurodes.

Each of these layers is "fully connected." This means that each neurode in the input layer sends its output to each neurode in the middle layer; similarly, each neurode in the middle layer sends its output to each neurode in the output layer. The output-layer neurodes' outputs are what we monitor to see the network's response to a pattern presented to the input layer. This is a reasonably extravagant

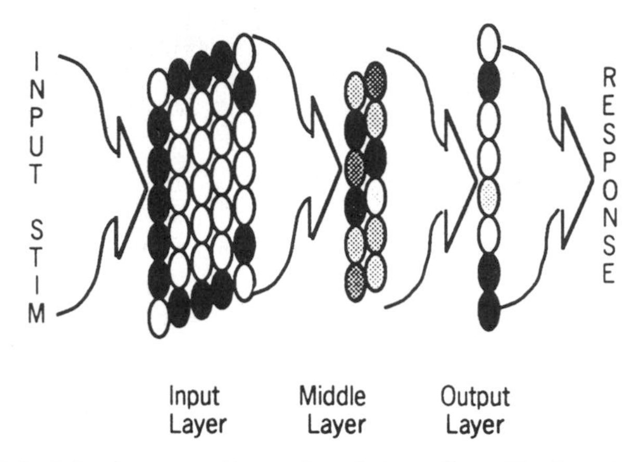

Figure 6.1 A backpropagation network in action. The input pattern is an image of the letter C*; the output pattern after training is the binary ASCII code for the letter, 0100 0011. This network has almost learned the correct output for* C.

connection scheme. A moment's thought shows that if each of the 35 neurodes in the input layer connects to the 12 neurodes in the middle layer, there must be 35 × 12 or 420 connections between the two layers. Similarly, the connections from each of the 12 middle-layer neurodes to each of the 8 output-layer neurodes consist of 12 × 8 or 96 connections. Thus, this network consists of 35 + 12 + 8, or 55 neurodes, and 420 + 96, or 516 connections.

Since the network begins knowing nothing, the connection strengths between the various neurodes are set initially to different small, random values. With such a start, it is to be expected that the network's initial response to the image of a letter is wrong.

Suppose the network is trained only on the letters *A, B,* and *C*. The training procedure is straightforward. First an image of the letter *A* is presented to the input side of the network. The black dots in the image act as stimuli to the corresponding neurodes in this layer, causing them to become active; the input-layer neurodes that correspond to the white part of the image, such as the triangular "hole" above the *A*'s crossbar, do not become active. In effect, the input image is reproduced in the pattern of active neurodes in the input layer; a snapshot of the layer's activity thus is also a snapshot of the input pattern presented to the network. These active neurodes in the input layer generate a signal proportional to the strength of their activity (i.e., how dark their corresponding dot is in the image) that is transmitted to all the neurodes in the middle layer over the variously weighted connections.

Even though each middle-layer neurode receives inputs from the entire input layer, each one actually sees a unique net input signal strength because of the unique, random weights associated with each connection. For example, the same output signal from a particular input-layer neurode might have a net received value in the middle layer that could be any real-numbered value (either positive or negative), based on the value of the connection strengths between the input and middle layers. Generally the net signal seen at the receiving neurode is calculated simply by multiplying the incoming signal by the strength of its connection to each middle-layer neurode. This allows a positive (excitatory) signal from the input layer to be received at a particular neurode as a net negative (inhibitory) signal, simply because the connection weight over that path happens to be negative.

In any event, the neurodes in the middle layer receive stimuli from the input layer that are quite different from each other, even though the signals start out the same because each middle-layer neurode has its own unique collection of connection weights. Because of this scrambling, the neurodes in this layer have distinctly different responses to the overall stimulus from the input layer. These responses cause some of the middle-layer neurodes to themselves become active enough to send signals to the output-layer neurodes.

Just as with the middle-layer neurodes, the output-layer neurodes also receive differing net stimuli, and for exactly the same reason. The varied, random weights on the connections between the middle and input layers assure that it is unlikely that any two neurodes in the output layer receive exactly the same total stimulus. Of course, since all these connection weights start with random values, we can imagine that some random collection of neurodes in the output layer become active. The random activity is the network's response to this presentation of the image of the letter *A*.

Unfortunately, such a random response is not at all what we want the network to do when presented with an image of *A*. We really want the network to make the second from the top and the bottom output-layer neurodes very active (to make the binary pattern 0100 0001), and suppress all (or most) activity from the other neurodes in the layer. Succinctly put, the network generates the wrong answer. This is where the supervised nature of the training procedure comes in. When the network generates its best guess of the correct answer, we present the output layer neurodes with the correct answer as we have defined it. In this case that means high activity for the second and eighth neurodes, and little or no activity in all other neurodes. Each neurode compares its actual output to what it was supposed to output according to this desired pattern. This comparison is generally a simple subtraction of the actual activity from the desired activity for each neurode, and it is called the error. The error can be either positive or negative, depending on whether the actual output is too low or too high. Using this error, connections to each output-layer neurode can be adjusted to improve its performance. A neurode's positive error increases the connection strengths to that neurode (makes them more positive or less negative) and thus tends to increase the neurode's output; a negative error decreases the connection strengths (makes them less positive or more negative) and thus tends to decrease the neurode's output.

The only problem with this scheme is that all the mistakes the output layer makes may not be due to the connections leading directly to it. It is also possible that a good deal of the error derives from errors in the signals sent by the middle-layer neurodes. Unfortunately, we don't have any way of knowing what those middle-layer signals should be; all we know is what the output layer is supposed to do, not what the middle layer is supposed to do. How can this be resolved?

In backpropagation networks a very neat solution is used. Before the output layer neurodes' connections are changed, their errors are propagated over these same connections back from the output layer to the middle layer, weighting them by the original connection strengths. (For stability reasons, an additional factor is also used, but that is not important for this discussion.) Each middle-layer neurode thus receives a net error back from the output layer that it can use to adjust the weights between it and the input-layer neurodes. Only after the errors are propagated back are the connection weights updated to reflect this new experience.

A backpropagation network thus learns through this two-step procedure: First an input pattern is presented, and the corresponding activity is propagated forward through the network, and then the corresponding error in network output is propagated backward from the output layer to the input layer, modifying the connection weights as it goes. Once this two-stage pass is complete, the next training image is presented, and the entire operation repeated.

Backpropagation training is not only supervised, it is also highly iterative. This means that even after the network has seen all 26 letters in the alphabet, it likely still cannot recognize them. Instead, it must see the letters many times, with each presentation of a letter involving this same two-stage training process to recognize them correctly. Depending on the problem, training a backpropagation network might involve anything from 10 or 15 passes, to a few hundred passes, to hundreds of thousands of passes through the entire set of training examples. Backpropagation training does require patience.

On the other hand, while it may be relatively slow to learn, backpropagation networks seem to be able to learn almost anything. They are highly forgiving, and easy to build (although if a computer simulation is used, they can chew up enormous amounts of memory and processor time). And they have been successfully trained in problems ranging from credit approval for mortgage loans to controlling a car on a freeway.

Does backpropagation model the way people and animals actually learn? The chances are that it does not in general. There are several problems with backpropagation as a model of biological learning that make it unlikely—although not impossible—as a biological process. Probably the biggest of these problems is the way errors are propagated. In the nervous system, connections between neurons are almost certainly one-way connections. That is, a connection from cell *A* to cell *B* cannot pass a signal from *B* back to *A*. If such a link exists, it must be on a separate connection, not the same one connecting *A* to *B*. Yet backpropagation works in large measure because the errors are weighted in their backward journey through the network by the very same weights that control the forward propagation of signals. If the forward weights and the backward weights are unrelated, there is little reason to believe that backpropagation's training scheme would work.

Backpropagation networks have other problems, too, including the fact that they work best in very small networks of a few hundred or fewer neurodes; the larger the network, and the larger the number of examples it has to learn, the more passes it takes for the network to learn and the more likely it is that it does not learn in any reasonable time. Yet biological networks tend to be huge, certainly orders of magnitude larger than a few hundred neurodes. While these arguments and others do not rule out backpropagation as a biological model, they do make it an unlikely candidate. On the other hand, most engineers have found that, biological model or not, backpropagation works extremely well for a very wide variety of problems. Almost certainly the android we build will have a few such backpropagation networks somewhere in its design.

But we still have to provide the android with learning that is much more similar to animal learning. We need the efficiency animals achieve in order to improve training performance. The only question now is, how does such learning take place in animals and people?

Animals exhibit several different kinds of learning. Probably the one most people have heard of is classical conditioning. This occurs when an animal or person learns to associate the appearance of one external event with another. For example, if every time that a child hears a particular bell ringing it is immediately followed by a frightening bolt of lightning, the child quickly associates the bell with the scary lightning and begins to show frightened behavior just from the sound of the bell. In effect, the child is now afraid of the bell, and not just the lightning.

The Russian psychologist Ivan Pavlov performed classic conditioning experiments, using dogs as the experimental animals. Dogs naturally salivate when shown a plate of food; they do not normally do so when they hear a bell. Pavlov found, however, that he could condition dogs to salivate at the bell's sound by consistently showing them the food and ringing the bell at the same time. The phrase "at the same time" needs to be taken with a grain of salt here; nevertheless, Pavlov demonstrated that the combination of the presence of a neutral stimulus like the bell's tone along with the sight of a highly emotional stimulus such as the plate of food could charge the neutral signal with emotional content.

Conditioning is thus the association of emotional content with a stimulus or event that was originally emotionally neutral. In the dog's case, the emotional reaction is the primitive hunger response of salivation; in the child afraid of lightning, the reaction is a frightened startle. Both occur—after conditioning—in response to the (initially) neutral stimulus of the sound of a bell.

Pavlov found that, generally speaking, conditioning occurred more effectively when the neutral stimulus (the bell) appeared somewhat before the emotional stimulus (the food). In other words, for greatest efficiency the two should appear not "at the same time" as indicated above, but rather sequentially, with a short delay between them. Upon consideration, this makes sense if we view conditioning as a survival tool for the animal. Suppose the idea is to train an animal to associate a particular scent with the appearance of a dangerous predator. If successful, such an association provides it with a warning that may be sufficient to avoid being eaten. Being able to predict the future, even in this simplistic sense, turns out to be a crucial survival skill. And because it is a very basic skill, it is found in extremely simple animals, ranging from people down to planaria (flatworms).

Animals from sea slugs and butterflies to snakes and people take advantage of this learning behavior by trying to force their prey or predators into an emotional response that ordinarily is associated with a much more fearsome organism. For example, an animal may mimic another that is more dangerous—or just more vile-tasting—in order to convince potential predators that it would be unwise to take a bite. And the old story of the wolf in sheep's clothing is an example of the predator mimicking the prey to increase its odds of a successful hunt. Mail-order houses that run sweepstakes very commonly use this technique, making their envelopes as similar as possible to those of government agencies such as the IRS, all in order to generate the fear response and inspire the recipient to open the envelope.

Animals also learn through a second training procedure, called operant conditioning. This means that they learn by observing the consequences of actions. There are two major kinds of consequences, positive ones and negative ones. Positive reinforcement means that the animal receives something pleasant or desirable when it (by accident or design) performs a particular action. Negative reinforcement means that the animal escapes something negative by its action. The distinction is key: Positive reinforcement is like a puppy receiving praise for retrieving a stick; negative reinforcement is like that same puppy running away from a larger, meaner dog down the street.

Negative reinforcement is not the same as punishment; it is rather the successful avoidance of something unpleasant. Because Johnny crosses to the other side of the street, he avoids being beaten up by the bigger Bobby. Johnny is not punished, but he does learn that he can successfully avoid the negative consequence of a black eye by maintaining a careful distance from Bobby. Johnny's successful handling of this situation is negative reinforcement. In contrast, if Johnny insists on staying on the same side of the street, he is surely subjected to a losing fight. In that case, the punishment is the black eye and bruises sustained because he has not successfully kept the peace.

The key difference between classical conditioning and operant conditioning is that the responses an animal makes in classical conditioning are generally involuntary. The dog has little or no control over how much it salivates, for example. On the other hand, operant conditioning generally concerns voluntary behavior. A pigeon can choose to peck or not peck at a red button; Johnny can cross the street or not, or even turn around and go the other way. Such actions are within the control of the animal involved.

Both classical conditioning and operant conditioning have biological constraints on them. Psychologists have found that certain animal species are more likely to learn from visual cues than from taste cues; others are exactly the opposite. Birds are more visual creatures than rats, so they learn visual associations easier than others. Rats, on the other hand, are nocturnal so they don't depend on sight as strongly, and often eat in the dark; they depend more on scent and taste than on sight, and find it easier to learn such associations.

Animals have yet another key learning technique, and that is observational learning. This is learning by watching the experience of other animals. For an animal to learn by observation, it must first be paying attention to what is going on around it. If one chimp does not notice that another is using a convenient twig to entice a few insects out of a log for a tasty snack, then the first chimp cannot possibly learn to do it himself. Just as obviously, the animal has to remember what actions were performed, and it must be able to perform those actions physically itself. There is absolutely no point in trying to teach a chimp to pronounce the word "supercalifragilisticexpealidocious"—it only irritates the chimp and frustrates the trainer. Chimps lack the proper vocal equipment to pronounce it. (This is why most modern attempts to train apes or chimps to "speak" have concentrated on sign language rather than spoken languages. It is also why the peasant animal trainer mentioned at the beginning of this chapter is almost certainly doomed to failure.) And of course, the consequences of the action must generate either a positive reinforcement—like the chimp's mid-morning snack—or a negative reinforcement, such as the avoidance of a fight.

Observational learning thus requires the attention of the animal, the memory to retain what is observed, the physical ability to perform the actions, and a reason for performing them. It seems clear from all this that observational learning is very much a cognitive task; it requires much more mental effort than the knee-jerk reaction of classical conditioning, or even operant conditioning. The animal must be aware of its surroundings in a way not required for conditioning to occur. Furthermore, observational learning is more self-motivated than conditioning. In other words, the animal chooses whether or not to learn by determining the amount of mental effort it devotes to the process; this is not necessarily true in operant conditioning, and hardly true at all in classical conditioning.

For the moment consider only simple classical conditioning. Can we build a system that performs this kind of learning? The answer is yes, we can. But to understand how, we need to understand what learning means on a cellular level within an animal.

For a long time, psychologists puzzled over what physical changes in the brain account for learning in an animal. In 1949, Donald Hebb presented the first specific (although qualitative) statement of changes that occur in the cells during learning. To understand what Hebb proposed, consider two specific neurons, *A* and *B*, shown in Figure 6.2. Hebb stated that if neuron *A* generates an output signal that stimulates neuron *B* at time when *B*, for whatever other reason, just happens to be active, then the strength of the connection from *A* to *B* increases. In other words, the next time *A* tries to stimulate *B*, it finds *B* easier to prod than before, and more sensitive to *A*'s signals. This general statement is known as Hebb's Law, and it is the guiding force for a number of neural network learning algorithms today.

We can use Hebb's Law to build a system that exhibits classical conditioning. In the 1960s Stephen Grossberg of Boston University took Hebb's qualitative statement of cellular learning and turned it into a quantitative learning model called the outstar. In so doing, he

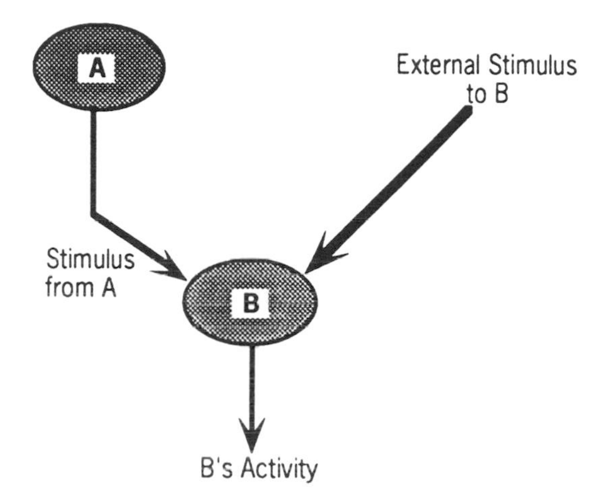

Figure 6.2 A simple, two-neuron system illustrating Hebb's Law. If A's stimulus arrives at B when B is active, perhaps because of some external stimulus, the weight on the connection from A *to* B *increases.*

realized that the model has to account for two key aspects of the network of neurons. It must express the changes in the activity level of each neuron at each instant in time, and it must define the connection strengths between neurons, particularly the changes in those connection strengths.

First, consider the neuron *B*'s current level of activity. Now, this activity could be caused by an outside stimulus—an eager researcher could be prodding the cell with an electric probe, for example, or a photon could be falling on the cell's photoreceptors if it is in the retina, or some similar outside effect could be occurring. In addition, stimulation could be caused by incoming signals from other neurons in the brain (such as neuron *A* in the figure). It is also possible that the cell is receiving no particular stimulus at all at this particular instant in time.

Grossberg generated a mathematical equation that describes how these three possible effects on the cell's activity interrelate. Essentially, the outstar's mathematical description has the property that increases in the cell's activity at any time are proportional to the sum of any external stimulus and the net weighted stimulus from other neurodes in the network. Because the cell is only concerned with currently arriving signals, the incoming stimuli from other neurons in the network at a particular instant in time may be from signals that were actually emitted by other neurons at some previous time. Like all other physical systems, signals within the brain can only travel at a finite speed, and take a measurable time to arrive at their destinations.

The net weighted stimulus is computed by taking the weighted sum of all the inputs *arriving* at this instant, not those being transmitted now. Finally, because the network has to be protected from random noise, each incoming signal must be at least as strong as some threshold value, or it does not contribute to the weighted sum. (Neurons in the brain occasionally fire randomly, even when they have no particular reason to do so.)

This takes care of any increase in the cell's activity at any particular instant in time, but it is impossible for any biological quantity to do nothing but increase; at some time it must decrease as well. Grossberg assumed that whenever the cell was not being actively stimulated, its activity should very rapidly decrease to near-zero, so that the cell could prepare for its next incoming stimulation.

While this covers the mathematics of each cell's activity, we are also interested in cellular learning, or how the connection strengths between neurodes change with time. The outstar model also mathematically describes Hebbian learning, or the changes in these connection weights. In this model, Grossberg assumed the simplest possible mathematical form that has the properties described by Hebb's Law. The simplest mathematical expression that expresses Hebb's law relates the change in each connection weight to the product of the incoming signal (weighted by the current connection weight) and the current activity of the receiving neurode. Consider this for a moment. Substantial learning (in other words, significant weight changes) can only occur when both the incoming weighted signal and the receiving cell's activity are strong; if the cell is stimulated at a time it is inactive, very little learning occurs. And, of course, if the cell is not stimulated over a particular connection, that connection experiences no learning no matter how active the cell is.

The outstar also provides a description of another uniquely biological phenomenon—forgetting. In this model, forgetting consists of the slow, gradual decay of the connection path strengths. Whenever a particular pathway is not being actively reinforced and used, the model assumes that it decays a tiny amount. Over the long term, infrequently used pathways, like infrequently recalled memories, gradually decay to near-zero strengths. It is very much a neural network version of "use it or lose it."

With the outstar we can build an artificial system that displays simple classical conditioning behavior. Imagine that we have a rectangular grid of neurodes based on the outstar model, each receiving two inputs, one from a single cell, *S*, and one from some external stimulus—like an electric probe—that we control. Suppose we consider the activity levels of the grid neurodes as black-and-white dots for a picture of the letter *A* as shown in Figure 6.3. At any time we can

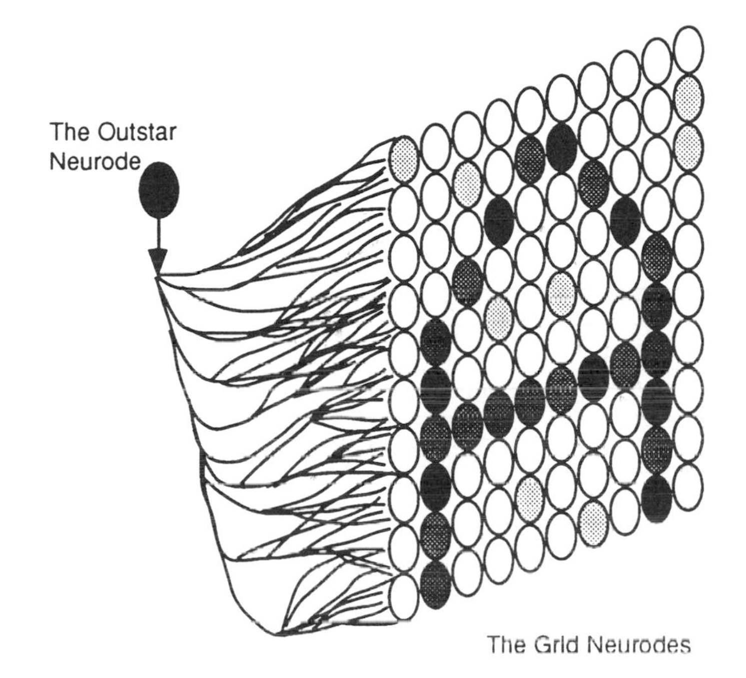

Figure 6.3 An example of a simple neural network that exhibits classical conditioning behavior. The activity of the outstar neurode is transmitted to all the grid neurodes along their individual connections. The network shown is only partially trained; thus, the grid's image has various grays instead of black and white.

use the external stimulus to force the grid neurodes into activity and display the picture we want. At the same time, we can also control (through some other stimulus) whether neurode *S* is actively stimulating the grid. Let's set up a training procedure that corresponds to classical conditioning. When we externally force the grid to reproduce a particular picture, it is like showing a dog a plate of food; the grid has no control over its response, and it displays the picture on demand. If we do this at the same time as we have neurode *S* stimulate the grid, it is like ringing the bell while we show the dog the food. The neutral stimulus from *S* (which originally cannot cause the grid to display anything but a random, "snow" pattern because the connections between it and the grid are random) takes the part of the neutral bell sound.

Consider one particular neurode in the grid that corresponds to a dark (highly active) spot on the picture. When it receives the stimulus from *S*, it is already highly active because of the external demand to reproduce the picture. The product of the high activity and the strong incoming stimulus means that the learning law causes the connection strength from neurode *S* to this grid neurode to increase strongly. If this is repeated enough times, a signal from *S*, by itself, is enough to cause this grid neurode to become very active. Thus, after enough repetitions, all the neurodes corresponding to dark spots on the picture generate dark spots solely on the basis of the *S*'s signal. No matter what the original weights are, sooner or later they grow strong enough to ensure that this happens; the only effect a different set of initial weights has is to determine how many repetitions are needed until the association is learned.

What about the grid neurodes that correspond to white dots (very low activity) in the picture of *A*? Do they learn not to fire when *S* stimulates them? Here we have to consider two possible cases, one where the original, random connection weight was weak, and one where it was very strong. If the original connection was a weak one, the stimulating neurode's signal is received very faintly by the grid neurode. Since there is little activity (because it corresponds to a white spot in the picture) at the time this faint stimulus is received, the product of the incoming signal and the current activity becomes even smaller and there is little or no weight change—except that the natural forgetting of the network weakens the connection strength slightly. Thus, this connection stays very weak and the grid neurode, even after training, does not activate in response to *S*'s stimulus. But what if the original random weight was such that any incoming stimulus from *S* is strong enough to make this grid neurode fire, even though it's not supposed to? In this case, we have to invoke the other term of the learning model, forgetting. Remember that in spite of the strong incoming signal, because the current activity of this grid cell is zero when it is received, there is still no weight increase. Furthermore, recall that the outstar model also forces weights that are not actively increasing to decrease through forgetting; the use-it-or-lose-it creed is vital here. This means that the connection experiences a net decrease in the strength of the connection through forgetting, just because there was no weight increase to counter it. Every training repetition causes a similar small decrease in the weight. If the training is repeated enough times, eventually the connection strength decreases to the point where *S* can no longer stimulate this grid neurode to fire.

After enough training repetitions ("enough" being determined by the original random weights, how fast learning is occurring, and how fast forgetting is taking place), the connections to grid neurodes that are supposed to be dark become very strong, and the connections to grid neurodes that are supposed to be white become very weak. Once this is the case, *S*'s stimulus by itself automatically causes the grid to reproduce the desired image. Just as ringing the bell caused the dogs to salivate after conditioning, having the originally neutral neurode stimulate the grid causes the desired picture to be displayed. This little network has truly modeled classical conditioning.

How good a model of classical conditioning is it? Well, the answer to that is "pretty good, but not great"—there are a few little glitches. In particular, the greatest problem is time. The outstar only works because it considers signals that arrive simultaneously with the current activity level. But if you recall, biological classical conditioning works best when the neutral stimulus (the bell) occurs before the emotional stimulus (the food) appears: Pavlov had to ring the bell first, before showing the dogs the food. In fact, in some cases, psychologists have shown that the best results are obtained when the neutral stimulus precedes the other by as much as several hours! The outstar simply cannot account for this at all. Furthermore, there are other technical details regarding classical conditioning that the outstar does not correctly model. Does this mean that all is lost? By no means, because it turns out that there is an even better version, called drive-reinforcement theory.

Drive-reinforcement theory was developed by Harry Klopf of Wright-Patterson Air Force Base, and it changes one of the basic assumptions of the outstar model. In essence, Klopf says that Grossberg's simple mathematical expression for Hebb's Law is wrong. The weight change between two neurons is not proportional to the product of the incoming signal and the current activity, but is proportional to the product of the current *changes* in them. This means that if an incoming stimulus is increasing at a time when the receiving cell's activity is increasing, the weight involved also increases. However, if the incoming stimulus is increasing at a time when the receiving cell is decreasing, then the weight decreases, instead of increasing (a positive stimulus change times a negative activity change gives a negative weight change). Furthermore, Klopf's system contends that the current weight change must be based on the net effect of the stimulation the cell has received over some past period of time. So it is not just what is happening now that matters, it is the history of what has happened to that cell over some time period in the past. (The length of this historical period obviously depends on the particular problem or system involved.) The oldest historical events have less effect than the most recent ones, but they do have an effect.

When Klopf's system (including a number of additional details not mentioned here) is built as an artificial neural network and its training performance compared to that of classical conditioning in animals, the results are astonishing. The drive-reinforcement system models classical conditioning exactly in nearly every detail. It is a triumph of psychological modeling that truly provides insight into animal learning.

From the perspective of building a trainable android, we can easily model the classical conditioning behavior of animals. With small variations on this, we can likely also model operant conditioning as well. With backpropagation we have additional training techniques that may or may not be biological, but allow us to effectively train systems to perform a wide variety of tasks. But this is not enough. All these systems require us to tell the android exactly what it is we want it to do. These are all supervised training schemes, just like the peasant teacher trying to train a horse to sing. We also need a system that can learn on its own just by observing the world around it; it must be able to perform observational learning. The next chapter takes up this challenge.

≧ 7 ≦

Discovering the Truth

Most people would sooner die than think; in fact, they do so.

Bertrand Russell

Little in life is certain, yet there is one characteristic that we can count on: Nothing is more constant than change. The world around us is different every year from the years before, and every day from the days before. The environment we cope with so easily today can become a trap for us tomorrow—or next year. The training we receive as children must constantly be modified and adapted to the conditions of today. Yet we still manage to live and to prosper, both individually and as a species. How do we accomplish this?

It seems that our species has developed into one vastly skilled at adaptation. Our biological weapons such as teeth and nails are almost laughable compared to those of a lion or grizzly bear, yet we get by very well indeed. Human survivability is not so much one of greater intelligence—dolphins and whales, after all, are probably about as intelligent as we and yet they hover on the brink of extinction—as it is one of innovation. We have the capability to change, to create new solutions to new problems, and to implement those changes immediately, without waiting for the slow march of genetic mutations. Apparently we are one of the few species on earth that has invented the notion of cultural innovation, and we alone have made it into a compulsion.

Innovation seems a hallmark of our kind, and is thus one characteristic we must include in the repertoire of an artificial human. This implies two kinds of skills. The first is the ability to learn from direct experience, or observations of the direct experience of others, even when there is no obvious tutor to explain the "right" answer to a problem. The other is skill at problem solving. This chapter addresses the first of these; problem solving is deferred to the next chapter.

Learning without a tutor happens when unexpected events—either good or bad—occur during the lifetime of an organism. Some of these events are deemed important enough that they result in learning. In an android this kind of untutored learning can happen when it is operating in the natural environment, but it can also occur while it is still in training. In fact, an understanding of animal development leads us to believe that such "untrained training" may be a vital step in building the android itself.

Animals and people experience this kind of untutored learning throughout their lives, but it is especially prevalent in infancy. A newborn baby has all the visual apparatus necessary to see the world around it; her eyes and optic nerves function perfectly well. Yet a newborn cannot see in the sense that an adult sees; or, rather, the newborn sees, but she does not perceive what she is viewing. The baby must learn to understand the confusing hodgepodge of stimuli that bombard her. She must make some kind of reasonable order out of the chaotic sensations so that she can eventually understand that one kind of stimulus means a flat surface, another means a sharp edge, and another means the warm, comforting presence of her mother. Is there any way the child's parents can teach the baby how to see? No. All parents can do is offer a rich set of visual stimuli to encourage the baby to develop her eyesight properly. Thus, they hang colorful mobiles over the crib, provide pictures and murals on the walls, and give the baby brightly colored toys. If the child develops normally, after a few weeks the circuitry in her brain begins to make sense of these stimuli so that she can make accurate predictions about the objects and surfaces she sees.

That this process occurs after birth rather than during fetal development can be seen in an experiment performed with kittens. Newborn kittens do not open their eyes until a short time after birth, thus their brains do not receive significant visual stimulation until that time. If a kitten is reared in an environment in which every object and surface is painted with vertical stripes, it appears to develop normally. However, if that kitten is later placed in an environment containing only objects painted with horizontal stripes, the kitten experiences severe problems. In effect, it is now blind. Its visual system has not learned to interpret horizontal lines, only vertical ones, and it cannot understand what it sees. There is a critical period in the kitten's life during which it needs to be exposed to a variety of visual stimuli. Failure to experience a sufficient variety of sensations during this brief period results in a cat that can never overcome the deficiencies of its upbringing. The kitten in the experiment may never learn to understand horizontal lines, even if it spends the rest of its life in that environment. Its effectiveness at overcoming its earlier experience depends on how long it was "horizontally deprived" and on its age when moved to more stimulating surroundings.

We now know that it is not just the visual system that must organize itself after birth. There is now very good evidence that at birth any human child has the potential to learn any human language. All children begin with the brain circuitry to perceive all possible human speech patterns and sounds, and if exposed to them before the age of about 3 or 4 years, all will be able to learn any language with the unaccented skill of a native speaker. Once a child passes that age, however, the speech sounds (or phonemes) that she has not been sufficiently exposed to become less perceptible to her, while the processing of phonemes in her native tongue become strengthened. By the time the child is in school, she will have lost the ability to even perceive the sounds of languages other than those of her own language. And of course an adult is permanently locked into the vocalization pattern of those languages learned at an early age.

This does not mean that an adult cannot learn a foreign language; it means that unless the foreign tongue uses the same phonemes as his native tongue, an adult cannot learn to speak or understand the foreign language like a native speaker. This, by the way, is probably the root of the problem many Japanese speakers have with trying to learn the "l" and "r" sounds of English: Unless they were exposed to these phonemes as very young children, their brains literally lack the wiring necessary to distinguish between them. Native English speakers have a similar problem with certain sounds in languages like Chinese or Japanese, or even the glottal "r" sound of French. Intriguing experiments that appear to confirm this have been done with recordings of certain Native American languages. Some phonemes in these languages are not found in Western European languages like English or French. A researcher in Canada has taken tape recordings of different words that sound absolutely identical to a native-English speaker and presented them to children in a special laboratory setup. Each child was placed in a darkened room into which repetitive recordings of Native American words were piped. The children learned to look to one side when the words over the speaker changed because immediately after the word change a bright, noisy toy was briefly spotlighted. Once trained, each child listened to a tape of two identical-sounding words, words that an adult English-speaker cannot distinguish. Invariably, very young children had no difficulty in perceiving the difference between the two words and thus predicting when the toy would light up. Children a few years older, however, were unable to distinguish them, and failed to correctly predict the toy's appearance. Apparently humans, like kittens, have critical periods during which the brain develops essential skills.

Are such critical periods a weakness in the formation of the brain? After all, they imply that no person can ever learn to speak all human languages like a native. While this is a limitation, it is probably not a vital one. Every child is born with far more brain circuitry than is needed. During such critical development times, it is likely that key areas of the brain process language and speech are specializing and dumping redundant circuits and connections. While the result of this reorganization is that the child cannot speak every langauge equally well, it does allow her to speak her native tongue much better. In effect, the child's brain organizes itself to become an expert processor of English (or whatever her native tongue) rather than a good processor of language in general. And clearly, this skill is entirely an untaught one; we have no way of instructing the child's brain on how to do it.

There are other examples of such self-organization in the brain. Scientists know that the brain contains a large number of maps, called topology maps. These maps are sections of the cortex that are organized in accordance with the kinds of data they process. For example, in the auditory system there are sections of the brain containing neurons that are sensitive to various frequencies of sound. There might, for example, be a neuron that is especially sensitive to sounds of 1500 Hz (one Hertz, abbreviated "Hz," refers to one vibrational cycle per second), and another that becomes very excited by sounds of 1700 Hz. The interesting characteristic of these neurons is that they are physically organized in either increasing or decreasing frequency, depending on the individual. For example, a neuron sensitive to 1500 Hz sounds might lie between neurons sensitive to 1400 Hz and 1600 Hz. What is even more intriguing is the fact that at birth these brain cells are randomly located. Part of the post-natal development of the brain consists of lining these neurons up in order.

Sound isn't the only stimulus that causes the brain to reorganize. There are topology maps for the sense of touch that map nerve endings from the hand onto the brain, and even topology maps for location. A rat trained to run a maze has been shown to have neurons that fire when the rat is at a particular location in the maze; furthermore, these neurons appear to be physically organized like the maze itself. Unless we endorse the notion that this rat was genetically prewired to learn this specific maze, we have to accept that the process of learning the maze somehow physically changes the organization of neurons in its brain.

Topology maps in the brain are thus a known feature, as is the notion that they form after birth. But how exactly do they form? What kind of learning is it that models the physical structure of the brain on the data it processes? Scientists may have come up with a partial answer to this puzzle.

Teuvo Kohonen is a Finnish scientist at the Helsinki Technological University in Espoo, Finland. For many years he has been one of the world's leading researchers in neural networks. In the early 1980s he performed some landmark research and created an artificial neural network that displays the ability to self-organize. This network is called the Kohonen feature map.

His neural network is one of the simplest that exhibits self-organization, and is shown in Figure 7.1. In its most elementary form, it is merely two layers of neurodes, an input layer and a "Kohonen" or "feature map" layer. The input layer distributes the received external pattern to all the neurodes in the Kohonen layer; that layer then categorizes the patterns based on stimulus pattern received and the modifiable connections between the two layers. Generally, the two layers of the network are fully connected, meaning that each Kohonen-layer neurode received an input from each input-layer neu-

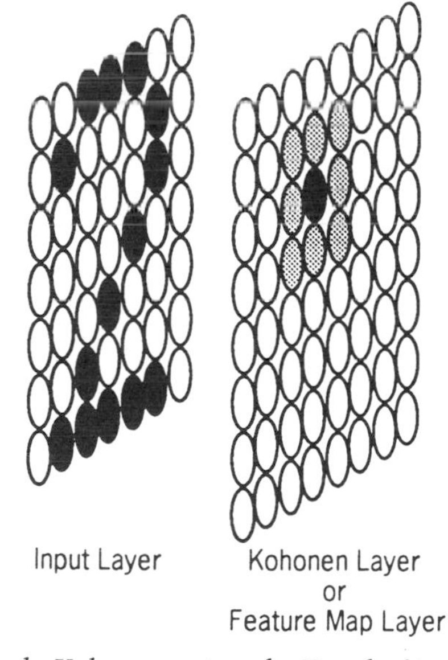

Figure 7.1 A simple Kohenen network. For clarity, the connections between the layers are not shown, but each input-layer neurode connects to each Kohonen-layer neurode. The winning neurode in the Kohonen layer is shown in black; its physical neighbors are shown in gray. Only the winner and its neighbors modify their weights in each training cycle; the only weights that are modified are the ones on the connections between the layers.

rode. The neurodes in the feature map not only have connections to the outside world, but also have significant connections within the feature map layer (not shown in the figure). These connections enable the network to implement lateral inhibition.

Lateral inhibition is the process by which a neurode attempting to generate an output signal interacts with other neurodes that are also attempting to output a signal. Each neurode's connections to the other neurodes in the layer are excitatory to its immediate neighbor neurodes, but inhibitory to those farther away. The effect of this is that each neurode tries to fire itself, tries to help its neighboring neurodes fire, and also tries to keep all other neurodes from firing. This sounds a bit strange at first, but it is actually a very effective means of allowing the neurodes in the layer to implement a self-regulating competition. In essence, the neurodes "fight it out" among themselves, and the result is that the single neurode with the strongest output signal survives to fire, and all others have their inputs squashed.

What does "strongest output signal" mean in this case? Each neurode in the Kohonen layer receives the complete input stimulus pattern to the network over the connections between the input layer and the Kohonen layer. Nevertheless, the pattern received by each Kohonen-layer neurode is modified by the individually weighted connections between the neurodes and the input layer, so that even though some external stimulus signals are transmitted along each set of connections, the total input received (and thus the net weighted input strength that determines the neurode's activation) is different for each Kohonen-layer neurode. The varying weights from connection to connection assure that this is true. Somewhere in this list of net weighted input signal strengths there is a single neurode that has the largest value; this neurode is the one that responds most strongly to the input pattern, and therefore generates the strongest output signal initially. This strong response causes it to be most effective at squashing all other neurodes' attempts to respond to this input pattern. In effect, this neurode has become the current "king of the hill." While this competition takes a little time to complete, it is self-regulating and always settles on the neurode with the strongest activity as the winner.

It is possible to predict in advance exactly which neurode in the network will win this competition. The input pattern consists of some number of signals, each of which arrives at a particular neurode over its own, separately weighted connection. If we compare the value of the connection weight to the value of the incoming signal for that connection, we can determine how closely the weights match the input pattern. The neurode that has a set of weights that most closely matches the set of signals in the input pattern is the winner.

Two characteristics make lateral inhibition important. The first is that it provides a means by which a network can determine for itself which neurode has the largest output signal. The network does not need an external, regulating mechanism to compare the outputs of its neurodes and make a decision; the network can determine the winner for itself, without recourse to any central control or intelligence. There is nothing magical or mysterious in this process, and it works very nicely indeed. The second reason lateral inhibition is important is that it is directly analogous to parts of the brain. In the visual system, for example, it is well known that excited neurons tend to inhibit the firing of their remote neighbors. In the brain, this is known as an "on center-off surround" architecture because when the center neurons turn "on," inhibitory links to their surrounding neurons turn them "off." Such biological consistency is usually a sure sign that an artificial neural system is likely to work well.

In a Kohonen feature map, an input stimulus presented to the network causes the neurode with the strongest response to fire; this alone does not provide a way for the network to reorganize its physical structure. The learning rules are what actually change the network. In most other neural networks, learning consists of changing all the weights on all the neurodes in the network for each training pattern. In the Kohonen feature map, however, learning only affects the winning neurode and its immediate neighbors—those neurodes that are its closest physical neighbors. The learning law used for this network is quite simple: For each input connection to a neurode undergoing weight adjustment, each weight is modified so that it is slightly closer in value to the input signal of the training pattern. Typically, the weight is adjusted by about 20 percent of the difference or less. Thus if the weight is originally 0.24 and the input signal is 0.74, then the weight increases by no more than about 0.10 (20 percent of the 0.50 difference between 0.74 and 0.24). If the input signal is smaller than the weight, the weight decreases in value; if it is larger than the weight, the weight strengthens. Each input connection of the winning neurode and its immediate neighbors adjusts according to this rule; the remaining neurodes' connections do not change.

It is easy to see how this learning rule works. The winning neurode is always that neurode whose pattern of weights are closest to the input signal pattern. Furthermore, the training procedure changes the weights so that the winner, and its immediate physical neighbors, have a pattern of weights that are even closer to that input signal. The next time this exact pattern is applied to the network—or the next time a similar pattern is applied—chances are very good that this same neurode will be the winner again, and that it (and its neighbors) will have the weights on their incoming connections nudged still closer to this input pattern. In effect, this particular neurode, although randomly selected with the first pattern presentation, has now become a filter that searches for similar input patterns. And because every time this winner adjusts its weights its neighbors also move their weights closer to this input pattern, they also become similar filters.

To make this more specific, suppose that the network is processing sound, and that the winner above reacted to a sound frequency of 1500 Hz. If the network is presented with 1500 Hz patterns several times, one particular neurode, say neurode *A*, slowly adjusts its weights (those between it and the input layer, but not those between it and other neurodes in the feature map layer) until they model a 1500 Hz signal very well. During this process, those neurodes physically closest to *A* (call them *B* and *C*) also adjust their weights closer to the 1500 Hz signal. Suppose now that the network is presented with a 1400 Hz signal. For this input pattern, neurode *A* is no longer the winner, but because its neighbors' weights have been dragged over near the 1500 Hz pattern, the chances are that one of them ends up as the winner. A few repetitions of the 1400 Hz pattern assures that this new winner, *B*, is solidified. At the same time, of course, *B*'s neighboring neurodes (*A* on one side, and another neurode, *D*, on the other) also modify their weights so that they are closer to the 1400 Hz signal. Although one of those neighbors is *A*, the small changes in its weights—remember the 20 percent maximum weight change—are not too significant, and *A* remains a good filter for 1500 Hz. (This is one of the reasons that each learning repetition only moves the weight a small percentage of the distance to the input signal.) If the network is now presented with a 1300 Hz signal, the winner will not be *A*, and it will not be *B*, but will likely be *D*, the neurode on the other side of *B*. Why? During *B*'s training *D* modified its weights so that it was nearly at 1400 Hz, an excellent position from which to become the winner for a 1300 Hz pattern. If this training procedure continues long enough, eventually the neurodes respond in an ordered fashion to decreasing sound frequencies.

The Kohonen feature map provides a fascinating and extremely simple model for the topology maps found in animals. Without recourse to any sort of central control or intelligence the feature map physically organizes itself and constructs similar maps to those found in the brain. For best results the training usually needs hundreds or thousands of pattern presentations, each making very small changes in the weights of the winner and its neighbors. Also, usually the frequencies are presented in mixed order, rather than all the 1500 Hz examples, then all 1400 Hz examples, and so on. While a bit tricky to implement correctly, the result is a network that is physically organized like the input data it is trained on, with adjacent neurodes responding to adjacent input patterns.

There is another benefit of this network as well: The clustering of weight patterns constitute a model of the likelihood that a similar input pattern has been seen by the network. For example, in the feature map example above, suppose that nearly half of the input stimulus patterns presented during training were in the range of 1400 to 1500 Hz. After proper training with such a collection of patterns, we would find that about half of the neurodes in the network had weight patterns corresponding to that same frequency range. There might be a neurode that responds most strongly to 1405 Hz, one that responds to 1415 Hz, and so on, so that half of the network's responsiveness was devoted to this small range of frequencies. The remaining half of the neurodes would have weight patterns spread over the rest of the audible range. This tells us that it is difficult to give a specific answer to the question of how accurately the network can distinguish between sound frequencies. For those sounds in the sensitive zone, 1400 Hz to 1500 Hz, the answer might be quite accurate indeed. On the other hand, for rarely seen frequencies, there might be only one or two neurodes that respond to broad frequency ranges covering 500 Hz or even 1000 Hz. Certainly, the more likely it is that an input pattern will be presented, the more accurately the network responds—and conversely, the less likely an input pattern, the less accurately the network responds. And as long as the network continues training (adjusting its weights), it continues to maintain an accurate model of the input pattern distribution, even if that distribution slowly changes in time. (If the distribution changes too abruptly, the network lags behind and is somewhat inaccurate until it has had a chance to catch up to the new input pattern distribution.)

The Kohonen feature map provides a way of categorizing input patterns. Notice that the categorization is an arbitrary one; the network chooses for itself which neurode corresponds to a given category, and even how dissimilar a pattern has to be to a category before being considered a separate example. On the other hand, the feature map creates the categorization entirely without a tutor to tell it what to do. It receives only the stimulus patterns as input, and not only provides a model for the biological topology maps, but also throws in a statistical model of the input distribution. Finally, it demonstrates that the biological on center-off surround architecture can be used to self-regulate a competition between neurodes, ensuring that the single neurode with the largest output signal is the winner, and that all other neurodes are prevented from firing. These are all critical achievements in neural network development.

Unfortunately, the simple Kohonen feature map network, while good enough for the kinds of raw sensory categorization we find in the early auditory system, is insufficient for the truly intelligent categorization that people perform constantly. What are its limitations in this area?

One problem is that it is rather finicky in the kinds of data that it can accept as input. In real life, an android must be able to take information as it flows from the unregulated, sometimes chaotic external world. The problem is that sensors that measure real-world values—like the number of photons striking a retinal cell, for example—can only report signal strengths as they are provided. A network like the Kohonen feature map that requires inputs to be in a restricted range and with a particular format will have difficulty unless those sensor readings are preprocessed by some intermediary before being presented to the network.

The whole concept of preprocessing a signal, whether it later goes to a neural network or to some other system or device, is an important one, and it is worthy of a brief diversion to discuss the issue. Neural networks, like any other processing system, always have a specific range of data values that they can accept; typical ranges are (−1 to +1), (−0.5 to +0.5), or (0.0 to 1.0), depending mostly on the preference of the developer. Signals can be converted to an appropriate range using any combination of four basic preprocessing techniques: scaling, normalization, filtering, and transforming.

The simplest of these is scaling. This merely means that the range of values of a signal is changed so that the minimum and maximum values lie within the acceptable range for the processing system. Scaling can be accomplished in several ways. The easiest method is linear scaling. Suppose, for example, that the input value is the outside temperature. A reasonable range of actual measured values might be from −40 degrees to +140 degrees on the Fahrenheit scale, thus covering a total of 180 degrees. Now suppose that this temperature reading is to be processed by a neural network that has a dynamic input range of −1.0 to +1.0 (a total range of 2 units). In order to make the measurement data match the input requirements for the network, the data values must be scaled so that they fall into this range. This is usually done using a simple linear scale like that shown in Figure 7.2. In this case a temperature measurement of 80°, being 120° from the bottom of the temperature range, is two-thirds of the way from the bottom to the top of the range; the scaled reading is similarly two-thirds of the way (1.33 units) from the bottom of the −1.0 to +1.0 range.

Scaling does not always have to be linear. Sometimes it is done nonlinearly. It is frequently helpful to use a "squashed" scale—one in

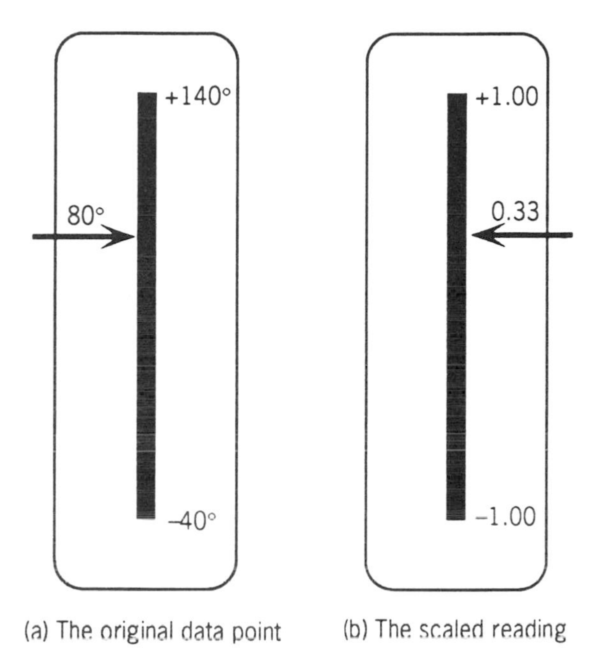

Figure 7.2 A temperature reading of 80° on a −40° to +140° scale converts to a measurement of 0.33 on a −1.0 to +1.0 scale.

which excessively large or small values are arbitrarily forced into a tiny range of scaled values—when a few measurements are significantly larger or smaller than most others. For example, if most temperature readings are in the range of 0° to 100°, we might assume that an 80° reading scales to about 0.6, or about 80 percent of the distance from −1.0 to +1.0.* Should an exceptional reading higher than 100° be measured, we can convert it to 1.0, the highest possible value in the range; similarly, a below-zero reading is converted to −1.0, the lowest possible value in the range. This technique is also useful in emphasizing important characteristics of an input stimulus while suppressing unimportant details. If the most interesting part of a signal occurs in a very restricted range, those values can be spread out so that they cover nearly the entire input range (−1 to +1, for example) of the system. Anything outside that range is just forced to the highest or lowest values as appropriate.

Sometimes scaling is not enough. Certain neural networks, such

*A measurement of 80° is 80 percent of the way from the bottom of the temperature scale (0°) to the top of the scale (100°). The total range from −1 to +1 is 2.0 units, and 80 percent of 2.0 is 1.6 units; thus, 80° scales to a reading of 0.6, which is 1.6 units from the bottom of the scale.

as the Kohonen feature map, require that their inputs be not only scaled but normalized. Normalization usually means that all input stimuli have a constant "length."

The concept of an input stimulus "length" needs explanation. Typically, a real-world problem requires that the system deal with the results of a number of measurements rather than the interpretation of only a single measurement value. (Usually such single-measurement problems are so simple that no special effort is required.) The solution system, of whatever type, generally needs to receive an assessment of the "state of the world" at a given time; that state typically can only be described by the current values of anywhere from several to several thousand different measurements. All these values typically must be presented to the network simultaneously. The individual measurements under such circumstances can vary widely in range, and usually the important characteristic is not the individual values of the measurements, but the overall pattern of the measurements. (Recall that neural networks react to an input stimulus pattern rather than any individual measurement within the input pattern.)

The collection of input signals presented to the network is organized into a consistent order. Suppose there are three measurements for a particular input pattern: temperature, air pressure, and humidity. The order must be consistent across all the patterns; the network cannot learn the patterns if one pattern has the mesurements ordered as "temperature, pressure, humidity" while another orders them as "pressure, humidity, temperature." It is possible to plot the input pattern as a point in "input-space" where the axes each represent one of the measurement values. The distance between this point in input-space and the origin (where all measurements are zero) is the length of the input pattern. Figure 7.3 illustrates this notion.

A pattern such as (temperature, pressure, humidity) typically has elements with very different units of measurement and value ranges; it is thus difficult to think of them as being part of the same unified whole. The way to get around this problem is to normalize the input patterns to a fixed length. This also has the happy result of removing the units of measurement so that the numbers become abstract values rather than inches of mercury or other units.

Normalization is usually performed by a simple, two-step procedure. First the length of input pattern is computed, usually by summing the squares of the individual input signals and taking the square root; this length is then divided into each individual signal-element of the input pattern. In effect this modifies the input pattern so that it has a fixed length of 1.0 and so that the individual pattern-elements have no particular measurement dimensions. After normalization, the in-

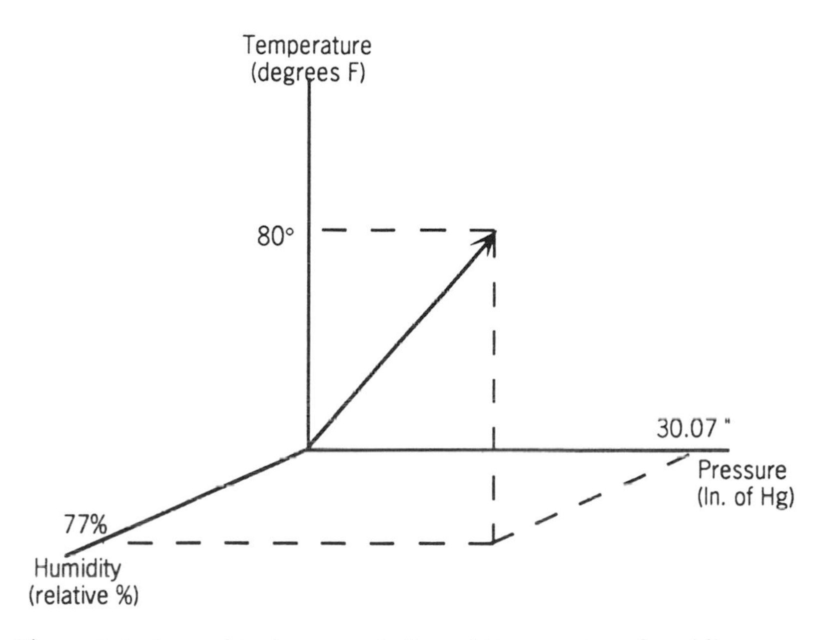

Figure 7.3 A graphical representation of (temperature-humidity-pressure) space. The arrow points to one specific input pattern with measurements of 80°, 77% relative humidity, and 30.07 inches of mercury. The length of this input pattern is the same as the length of the arrow.

put pattern's representation in input-space is still positioned in the same direction from the origin; only the length of the arrow pointing to the input pattern has changed to now be exactly one (dimensionless) unit long. Thus, a graph of the normalized pattern looks exactly like the one in Figure 7.3, except that the length of the arrow is now 1.0 and the three axes have no units of measurement associated with them.

This lack of units of measurement is important in a Kohonen feature map because the learning law involves subtracting the winning neurode's connection strengths from the input pattern presented. It is mathematically incorrect to subtract a unitless connection strength from a value that has units of measurement associated with it; in other words, you can't subtract an abstract number from a distance, or a temperature, or a pressure. Therefore, normalization of some sort is a requirement for this network.

Such simple Euclidean distance normalization is only one method that can be used; a variety of other mathematical techniques may be appropriate depending on the problem to be solved. Nevertheless, all normalization has the characteristic that the units of measurement are mathematically removed with the resulting input pattern having no specified units.

The third important preprocessing technique is filtering. Filtering is a process by which unimportant or distracting information is removed from an input value, leaving only the most important features for the system to process. Filtering can be done in many ways, and the specific technique used in a given case depends primarily on the input signals involved, as well as the desired conversion. Probably the simplest filter is a "band-pass" filter in which measurement values outside a specific "band," or range of values, are barred from entry to the system. These can be "low-pass" filters, which filter out everything but the lower end of the measurement range, or "high-pass" filters, which only allow passage to values at the upper end of the measurement range. Such filters are frequently applied in high-quality radios, cassette players, and stereos. When the high-pass filter is applied, bass tones are removed and the sounds generated are higher in pitch. When the low-pass filter is applied, treble tones are suppressed so that the sound quality is deeper and richer. If both are applied, only the middle range tones are allowed through.

Filtering is also one of the more common applications of simple neural networks. Networks make excellent noise filters because of their ability to reproduce relatively "pure" signals from garbled or noisy inputs. Noise in this context refers to superfluous measurement details that tend to disguise or even overwhelm an input signal with other meaningless signals. Frequently, noise consists of random variations in a measurement. A good everyday example of noise is static on a radio. In severe cases, the noise (static) tends to drown out the signal from the radio station, making it more difficult to comprehend the intended message.

The fourth important preprocessing technique is that of transforming the input pattern. Transformations, like normalization, generally operate on the complete pattern, rather than on individual measurements. Probably the most frequently used transformation for sensory information is the Fourier transform discussed in Chapter 2. It is particularly appropriate for vision and sound signals. A number of other specialized transforms can be used also, depending on the application.

While using appropriate preprocessing techniques assists with the problem of inappropriate input signals, other issues arise with the Kohonen feature map as an autonomously learning system. One major difficulty is that it provides only a fixed categorization of each input pattern. Once a given pattern falls into category *A* of a trained Kohonen network, it always falls into that category (unless substantial new training is applied to the network). While this might not sound like much of a drawback, consider how people categorize items.

Words and concepts are very slippery to people. We can consider the same object as good or bad, delightful or irritating, large or small, all solely on the basis of context. For example, many may consider an artificial sweetener to be less desirable than sugar or honey; others consider it essential to maintain a steady weight level. Even a particular individual modifies her categorization from "not as good as sugar" to "preferable to sugar" merely on the basis of whether she is on a diet at the moment. People seem to have an elastic notion of categories that changes from day to day and from minute to minute as circumstances and context change. Only children really believe that heros in white hats always do good deeds, and bad guys in black hats always do evil. The rest of us are painfully aware that the person we consider a good guy today may be caught with his hand in the cookie jar tomorrow; and that one who seems an all-out cad may also be exceptionally kind to children and an excellent parent. The realities of the shades of gray of our world insist that any android we build must also be able to keep up with these changing categories.

The Kohonen feature map also has the problem that it learns literally everything presented to it. Again, this might not sound like a serious problem, but consider its implications. Suppose you learned everything you see. That means you would memorize the nutritional content from the side panel of the box containing your morning cereal—and all the commercials you see. You would recall the names and faces of everyone you pass, even those seen casually on the street. You would store all the license plate numbers of all the cars that pass you on the freeway and memorize the brands and package sizes of every product on the supermarket shelves. And of course if you learned truly everything, it would include all the sounds you hear, all the textures you touch, all the odors you smell, and all the flavors you taste—even those of the burned toast from last summer and the nauseating concoction your little sister made when you were seven. If you truly learned everything you experience in the course of your life, you would clutter up your brain with so much trivia that there would be no possible way you could make sense of it.

But people don't actually learn in such an indiscriminate fashion; instead, we choose to store only information that is important or relevant to us. We don't remember all license plates, for example, only our own (if that one). We don't store every person's face we meet, only those we consider significant. We don't memorize all the products in the supermarket, only those we like or buy regularly. In all these ways and hundreds more, people exhibit an ability to select only critical information from the sensory cacophony that surrounds us, and to store only the most meaningful choices. An artificial person must also have this ability to "tune out" irrelevant and trivial details and concentrate only on those that matter.

The problem with this is that somehow people determine for themselves what is relevant and what is not. Granted, when taking a class or reading a book, the teacher or author provides strong clues to key points; yet even with this assistance people still make the final determination of relevancy within themselves—in the final analysis, no one can actually force anyone else to learn something. How do people determine importance?

One possible answer, suggested by Stephen Grossberg and Gail Carpenter of Boston University, is that important information is persistent in the sense that it is non-ephemeral. Unimportant details are transitory; they appear and disappear with little permanent effect. Important information, on the other hand, tends to either be present for significant periods of time, or to recur frequently. Thus the full moon, while only lasting a single night at a time, happens every 28 days or so. This made it an important event to early societies, and an enormous amount of information—and speculation—was gathered about this phenomenon. Similarly, if a new person appears in an elevator ride at work, it is insignificant and usually forgotten immediately; if that new person begins working on your project or department, he or she is no longer insignificant and will be carefully assessed. This notion of important information being that which persists can be used to provide an artificial system with another characteristic of human learning: attention.

People generally learn little or nothing unless they pay attention to the material presented to them. Attention provides us with the ability to focus our concentration on the critical information and ignore irrelevancies. You concentrate on an interesting speaker's words, for example, and ignore the hum of the air conditioner. Only if the background sensations are obtrusive or unusual do you find your concentration drifting away from the speaker to settle elsewhere. In a sense, attention is the tool you use that allows you to exercise judgment in what you learn. Thus, it appears that the android must have a means of paying attention to key issues, and ignoring, at least temporarily, other matters.

While attention seems critical for discriminating learning, it also imposes a limit. Few people can truly pay attention to more than one thing at a time. You often get around this by "time sharing" your attention: perhaps attending to a book for a few seconds, then to the television for a moment, then to the conversation around you. This process, by the way, is perfectly analogous to the way a multitasking digital computer works. It runs multiple programs "simultaneously" by allowing its single processing unit to spend only brief times on each program. Because digital computers process information so much faster than people do, these brief time "slices" appear to be the simultaneous execution of each program; in fact, each executes separately and sequentially. If we do not want the android to have tunnel vision, we must also make its attentional system "multitasking" so that it also can give reasonable amounts of consideration to several things at once.

Grossberg and Carpenter have developed a neural network that displays at least the rudiments of attention and re-categorization. (It is not capable of multitasking its attention.) This system is the adaptive resonance network, and it is useful to consider how it works to see how categorization and attention can be implemented. The discussion will necessarily be simplified compared to the actual network, and it only considers the first of Grossberg and Carpenter's adaptive resonance models, ART-1, but it illustrates several key ideas.

In discussing this network, we first have to understand that it cannot accept just any kind of input pattern; it must receive only binary patterns, consisting of 1s and 0s. This restriction is critical to the operation of the ART-1 network. (The more advanced—and much more complex—versions of the network, ART-2 and ART-3, drop this requirement.) Furthermore, the ART-1 network uses a special kind of storage node, called toggles or gated dipoles. These toggle nodes can be constructed from half a dozen or so ordinary neurodes in a particular structure, and they have a peculiar characteristic: When provided with a special signal called the global reset signal, an active toggle—one currently generating an output signal—becomes inactive and stays inactive for some period of time after the reset signal ceases. A toggle that receives the global reset command at a time when it is not currently outputting a signal does nothing, and can generate an output as soon as the reset signal is removed. In essence, the global reset signal prevents an output only in those toggles that are currently active, and this effect is persistent for some time in these nodes; all other toggles experience no long-term effect. For the remainder of this discussion, "node" in an adaptive resonance network refers to one of these toggles.

An adaptive resonance network has a complex operational sequence. Figure 7.4 illustrates the key parts of the network. Initially, the network is naive—it has seen no input patterns and therefore knows nothing. Suppose we present it with a particular binary pattern such as 1 0 1 1 0. This input pattern is received by three different sections of the network: the input layer, the reset system, and the gain control system. Initially, however, the input layer nodes (toggles) can do nothing. These nodes potentially can each receive inputs from

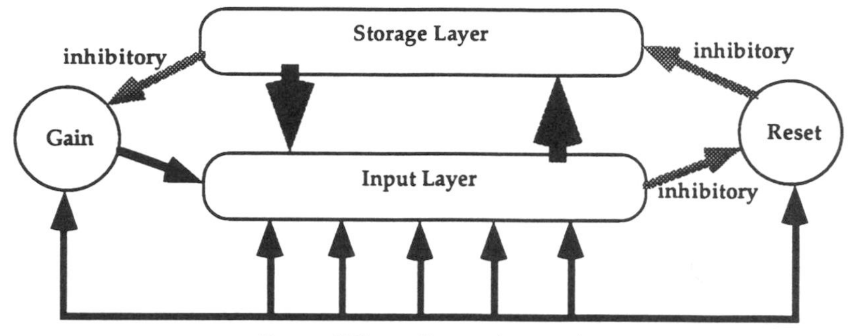

Figure 7.4 An adaptive resonance network. Gray connections are inhibitory; black ones are excitatory. Only the connections between the storage and input layers modify during training; all others do not learn.

three distinct sources: the external input pattern itself, the storage layer, and the gain control system. The network is set up so that each input-layer node must receive inputs from at least two of these potential sources in order to be able to fire. (This is called the "2/3 rule," since two out of the three sources of stimulation must be present to generate an output.) Since the network is quiescent when the input is first presented, neither the gain control nor the storage layer are active at the moment. Thus, the input layer is temporarily inactive.

The reset system also receives the input stimulus pattern directly. It consists of a single neurode (not a toggle, just a simple neurode) that receives signals over excitatory (positively weighted) connections from the external pattern and over inhibitory (negatively weighted) connections from the input layer. Initially, of course, it only receives the excitatory input pattern because the input layer has not yet fired. As a result, during this brief period it sees no inhibiting commands, and the reset system fires. Its output goes to the second layer of the network, the storage layer, where it acts as a global reset signal. The effect of this is to turn off any nodes (toggles) in the storage layer that are currently active; thus, the storage layer is cleared for possible recall action. Initially, of course, the storage layer is quiescent, so no nodes in it experience any long-term effect from the reset signal.

During this same brief period of time, the gain control system also receives the input pattern. The gain control has two inputs, one from the input pattern, and a massively inhibitory input from the storage layer. If any nodes at all in the storage layer are active, the inhibition is so strong that the gain control's activity is completely suppressed. Happily, however, the reset system has just shut down the storage layer's outputs, if there were any. This allows the input pattern to excite the gain control into activity. This gain signal is transmitted to all nodes in the input layer, where it acts as the second source of stimulation to those nodes. The input layer nodes are now able to fire.

Which nodes in the input layer will fire? Because of the 2/3 rule, only those nodes that receive two distinct sources of stimulation can produce an output. This means that only those nodes that correspond to the 1s in the input pattern receive an external stimulus as well as the gain stimulus; thus, the input layer faithfully reproduces the binary pattern of the external stimulus. The input layer's activity pattern, then, is the same as the input pattern: 1 0 1 1 0. This pattern of activity is transmitted to two places, the reset system and the storage layer.

At the reset system, the input layer's signals are received over inhibitory connections; the external stimulus is received over excitatory connections. But the two patterns are identical, as we have just seen, so they exactly cancel each other out, meaning that the reset system now has a net stimulus of zero. As a result, the receipt of the input layer's signals to the reset system shuts off the global reset signal to the storage layer.

At the storage layer, the global reset is now off, and there is an incoming pattern from the input layer. This pattern is called the bottom-up pattern (from the fact that the input layer is generally drawn below the storage layer in diagrams), and is transmitted to the storage layer over a series of modifiable connections. These connections, along with the corresponding ones that link the output of the storage layer to the input layer, are the only ones in the entire network that have adjustable weights. The two layers are fully connected, so each node in the storage layer receives the entire pattern transmitted by the input layer. As the 1 0 1 1 0 pattern is transmitted to the storage layer, it is altered by its passage across these weights, so that each node in the storage layer sees a somewhat different version of the pattern, thus generating somewhat different responses to it among the nodes.

The storage layer nodes now make use of a similar lateral inhibition architecture to that found in the Kohonen feature map to compete among themselves and determine the single node that best matches the bottom-up pattern. This single node, *A*, sends its output signal to two places: back down to the input layer and to the gain control system.

Recall that the gain control system has a very strong set of inhibitory connections from the nodes in the storage layer; receiving a single signal from any node in the storage layer is sufficient to suppress the gain control's output completely. Thus the gain control shuts off, and the second signal to the input layer that allowed it to generate an output in the first place is removed.

At the same time as the gain control's signal is removed from the input layer, however, the storage layer's signal arrives there. Just as the bottom-up pattern passes through synapses and is modified in the transmission to the storage layer, this signal also passes through the top-down synapses (again, so named because the storage layer is normally drawn above the input layer), and is similarly modified. This top-down signal will try to activate some of the nodes in the input layer. It will fail in general. Why? Because only those nodes that are still receiving a stimulus from the external input pattern can overcome the 2/3 rule and fire. Thus, only the first, third, and fourth input-layer nodes have any prospects of firing now because the external pattern has only these elements active. If these nodes are among those that the top-down pattern tries to excite, then they will become active and fire; if any is not, then that node will stop firing even though the external input is still trying to excite it.

Suppose that the nodes activated by the top-down pattern are nodes one, three, and four. That means that the pattern generated by the top-down pattern matches the external input; in effect, the network *recognized* the input pattern. It categorized this input pattern as belonging to category *A* (because node *A* in the storage layer successfully matched it). If this is the case, then the output of the input layer remains stable, it continues to match exactly the external input pattern, the reset system remains off, and the network has achieved a state called adaptive resonance. In an ART network, this is the next best state to Nirvana.

Suppose, however, that the nodes activated by the top-down pattern do not include one of the input pattern nodes, say node three. Node three cannot continue firing because it is receiving only one source of input, that of the external pattern. For the same reason, if the storage layer's pattern tries to excite node two (not part of the original input pattern) it fails, because node two receives only one source of input, that of the top-down pattern. Thus, from having nodes one, three, and four fire, the input layer's activity is reduced so that only nodes one and four now fire. This means that the reset system receives three excitatory signals (nodes one, three, and four in the external input stimulus) and only two inhibitory signals (nodes one and four in the input layer); thus, the reset system is no longer prevented from firing. It does so, and the resulting global reset signal shuts off node *A* in the storage layer.

Consider the effect of this: Node *A* has been shut off and will remain off for some time. No other node is currently active in that layer because *A* had squashed all competing nodes (and, anyway, if there had been any other nodes active, the global reset signal would have turned them off as well as *A*). Thus, there is no longer a strong inhibitory signal transmitted to the gain control; this turns the gain signal back on for the input layer, and allows the input layer to once again reproduce the original external 1 0 1 1 0 pattern. This of course shuts the reset system back off again, allowing most nodes in the storage layer (all but node *A*, which experiences persistent repression from the global reset signal) to become active. The input pattern is once again transmitted as a bottom-up pattern to the storage layer so an attempted match can be found. But there are now two distinct differences from before. First, node *A* is still repressed from firing because of the persistent effect of the global reset signal. Second, the weights on the connections between the input and storage layers are now slightly different than before. Every time a bottom-up or top-down pattern is transmitted across the links between the layers, the weights are adjusted slightly. This means that even though the bottom-up pattern starts off from the input layer exactly as it did before, by the time it crosses the synaptic junctions and arrives at the storage layer it is a little bit different. So the likelihood is that whichever node in the storage layer tries to match the bottom-up pattern will do a bit better than *A*'s attempt.

From this point, the network operates just as before: The storage layer nodes compete to determine a new category node, *B* (*A* being excluded from the competition). The winner shuts off the gain control and tries to generate a matching top-down pattern in the input layer. If it matches, the reset system stays off and the network reaches adaptive resonance; if it doesn't match, the total number of active nodes in the input layer decreases, the reset system comes back on, shutting off this new category, which permits the gain control to turn on, and the input layer to again send the bottom-up pattern to the storage layer. (This time, of course, both *A* and *B* are excluded from the competition.) Each attempt to match is done with weights that are slightly different from the time before, allowing many nodes an opportunity to be the best match.

Eventually, if the external stimulus persists, the network is guaranteed to either match the pattern, or learn it as a new category because the weights between the input and storage layers have changed enough that a good match is found. Notice that the adaptive resonance network learns only patterns that persist long enough for this rather complex procedure to occur. If at any step of the way the external pattern disappears, the learning process stops. If the measure of importance of a stimulus pattern is that it persists long enough to be noticed and learned, then the adaptive resonance network truly implements the quality of learning only what is important—and not having to be told what that means.

There is one other important aspect of the adaptive resonance network not yet discussed, and it has to do with the reset system. So far, the assumption has been that the inhibitory signals from the input layer must exactly match the actual input pattern to keep the reset system from sending out its global reset signal to the storage layer. What if the reset neurode had a threshold, such that mismatches that were smaller than this threshold were insufficient to cause it to fire? Suppose the input and external patterns had a total of 100 nodes, rather than the simple 5-node pattern used in the previous example. A threshold could be set in the reset neurode so that if 99 of the 100 nodes matched that was good enough to prevent it from firing; or we could set it at 90 nodes, or even 80 nodes. This threshold corresponds to the vigilance parameter, and it determines how closely the top-down category pattern has to match the actual input. Because of its definition, a high vigilance implies that the patterns must match very closely, and thus the threshold on the neurode is quite low—even a few mismatching nodes cause reset to happen. A low vigilance implies that any fairly close match will do, and thus the threshold is quite high—many nodes must mismatch before reset occurs.

Adjustment of the vigilance parameter also permits the network to change categorizations, just as people do. For example, suppose the network had many patterns to classify, say, 50 pictures of different flowers. If the vigilance parameter is very high, each of the 50 pictures might be placed into unique categories; if it is moderately high, the 50 flowers might end up in 10 or 15 categories; and if very low, all 50 might be classified in one global group. Similarly, people could classify them as 50 distinct species, or as 10 or 15 groups (daisy-like, rose-like, lily-like, and so on), or all as members of a single category, flowers. There is nothing intrinsically right or wrong about any of these classifications; it simply depends on how picky (how vigilant) the categorizer is in distinguishing differences between the examples.

There is another part of the ART network not illustrated in Figure 7.4. This is a second gain control that excites the storage layer rather than the input layer. This second gain control has the effect of forcing the network to pay attention to a particular input pattern. In effect, it tells the storage layer to "listen up" because something of interest is happening. This final addition provides the last key needed to model biological observational learning.

Does the adaptive resonance network solve the problem of providing the android with attentional and re-categorization capabilities? Unfortunately, the answer must be no. First, while adjusting the vigilance parameter (and other parameters) of the network does provide limited categorization and attentional control, this adjustment is not done automatically—the network designer must set the parameter appropriately for the problem at hand. No truly robust or adequate means of providing automatic vigilance control has yet been devised.

Second, the adaptive resonance network described works only because its input patterns are binary; if real numbers are substituted, it does not work at all. An android cannot expect to have its input predigested into convenient binary packages. The demands of the real world are too great to allow this in the final system; it must be able to handle data in whatever form it becomes available. And while more advanced versions of adaptive resonance have been developed that can process gray-scale inputs, they tend to be less reliable, less robust, and far more complex than the original adaptive resonance network.

Finally, there is a fatal flaw in the design of the adaptive resonance network that must be corrected before it can be used in an android: It uses "grandmother cells" for storage. The brain cannot use single cells to store complex patterns. Unfortunately, the adaptive resonance network does use single nodes in the storage layer to store complex categories. (Recall that the competitive architecture of the storage layer results in only a single node representing each category.) In a complex being like an android, it would be intolerable if the failure of a single node or, worse, a single connection, caused the loss of entire categories of information. The memory system must be robust; it must be able to handle a reasonable amount of hardware failure and still function well. A grandmother cell scheme simply will not do.

There is another problem with grandmother cell storage that may not be apparent. Storage of hierarchical concepts absolutely demands multi-cell representations. For example, suppose I want to create a memory representation of the orchid tree in my backyard. To gain the full measure of the concept, there must be a general representation of "tree," a secondary representation of "orchid tree" that contains the pattern for tree modified for this specific kind of tree, and a specific representation of "orchid tree in my backyard" that designates one instance of object orchid tree from the general category tree, expanded to add the notion that it is located in my backyard. The more specific the object or concept, the more nodes needed to provide a full representation. On the other hand, very general concepts and categories can often be represented by only a few nodes. When we use grandmother cell storage schemes, we lose the ability to provide a rich expression of such hierarchical categories. In the real world, it is likely that such fundamental relationships will be essential in an android.

While existing networks have not yet solved the problem of autonomous learning completely, great progress has been made. For example, some researchers, taking note of the large number of topology maps in the brain, have recently begun linking multiple neural network topology maps (similar to the Kohonen feature maps) to see if better autonomy results. Their conclusions are preliminary as yet, but very encouraging, and this may prove to be another major step in developing systems that truly learn independently. In the meantime, development of neural networks—or other devices—that truly learn without a teacher must continue before we can build a robust, complex system.

Not all learning takes place completely without a teacher, however. It is entirely possible to build a system that knowingly applies problem solving and other learning techniques to its experiences. The next chapter considers how such systems are constructed.

≧ **8** ≦

It's a Puzzlement

> **When I saw a sign on the freeway that said,**
> **"Los Angeles 445 miles,"**
> **I said to myself,**
> **"I've got to get out of this lane."**
>
> **Franklyn Ajaye**

Once a researcher in animal intelligence tried to test a chimpanzee's ability to solve a problem. A banana was hung from the ceiling of the test chamber, well out of the chimp's reach. The chamber was otherwise empty except for a long pole. The researcher wanted to see if the chimpanzee was smart enough to use the pole as a tool to knock the banana down. The result of the test was both unexpected and enlightening. The chimpanzee was brought into the room and the researcher watched it carefully to see what would happen. The chimp noticed the banana and apparently concluded that the fruit was too high to reach. The chimpanzee quickly picked up the pole, and stood it vertically on the floor just beneath the banana. The chimp then quickly climbed the pole, plucked the banana from the ceiling, and jumped down to the floor before the pole could fall over. The problem solution was creative, unique, and not at all what a less agile human could have predicted.

Androids will be faced with any number of problems to solve on a daily basis. If androids are to be useful in ordinary human environments, they will have to deal with these problems without constantly calling for assistance from a nearby human. Like the chimp, they will have to be good problem-solvers.

Problem solving involves both a system of reasoning and information about which the system can reason. Without some amount of knowledge about the problem to be solved and the environment in which the problem exists, it is impossible to work out a solution. Thus, in considering how an intelligent android might be able to solve problems, we must consider both how information is represented in the system, as well as how that information is used to work out a solution to a problem. Most often, the collection of facts that an AI system knows is called a knowledge base because it holds information in a specialized format. As a result, problem-solving programs frequently are called knowledge-based systems.

Consider one simple example of a problem-solving system. Suppose that you want to travel from New York City to Los Angeles. You may have a knowledge-base that has information on many kinds of travel techniques, such as auto travel, bus travel, train travel, and plane travel. How can you construct a solution that will get you to your destination?

Probably the most obvious solution technique is to begin by considering all the possible ways of traveling out of New York City. At the end of every time increment—say, an hour—of travel along each travel-path you evaluate the progress you have made toward Los Angeles. On the two or three most promising paths you continue to search all possible travel solutions from those various points, and so on. More specifically, after one hour of travel, you might have achieved the following locations:

1. By cab: at the airport
2. By bus: at the bus station waiting for a bus
3. By car: driving westbound, some distance from the starting point
4. By bicycle: five miles west of your starting point
5. By plane: not available from your house; however, see the taxicab's travel progress
6. By train: on the train, just leaving the station

From this list you might select numbers 3, 5, and 6 as the most promising future paths. For each of these possibilities you then extend the travel along that path for another time-step (an hour in this example). The result of this might be the following:

3. By car: driving westbound
5. By train: riding westbound
6. By plane: on an airplane, ready for takeoff on a westbound flight.

This step-wise process then continues until you find a path that reaches your destination in L.A. This process of searching for a solution beginning at the current state and working toward the goal state is called forward chaining.

In general, forward chaining is a process by which the current state of the world is changed by modifying one or more operators (such as "take a cab to the airport" in the example above). That operation changes the state of the world so that a new operator can be applied to the system. Ideally, each operator will reduce the distance to the goal position so that the eventual result is that the goal is achieved.

There are a couple of obvious difficulties with forward chaining for this kind of problem, however. To correctly reason forward, you must consider all possible next-steps at each step along the path. This means, for example, that you must at least consider driving the car over all possible paths departing from your initial position, including taking roads that lead east, north, south, and west. The reason this is essential is to avoid the infamous hill-descent problem mentioned earlier: It may be necessary to go east temporarily in order to efficiently travel west. This might happen if you have to travel east to reach the on-ramp for the westbound lanes of the freeway, for instance. The result of this is that the number of possible paths at the beginning of your search may be huge, and it may not be obvious until you have searched some distance into the paths which ones can be discarded as unfruitful.

Let's try to solve this problem again. This time, however, we can narrow the search by starting at the destination and working backwards from there to the initial state. This technique is called backward chaining.

In backward chaining, you first consider all the possible procedures that might result in reaching your destination. These might include taking a cab to the address you want, or driving there in a car, or taking a bus to the address. From there, you consider what procedure you might use to get to the point where you can take the cab, car, or bus to that address, and so on. Once you have that list, you continue backwards, always trying to find a procedure that will connect to the step you just made, but that might originate at a location nearer to your actual starting position in New York.

In general, backward chaining is the opposite of forward chaining. In backward chaining the system tries to find operators to apply to the goal state that result in the current state. It is the equivalent of starting at the finish point of a maze and trying to find your way to the beginning.

Forward chaining and backward chaining are complementary reasoning techniques, and each is useful for different kinds of problems. Forward chaining is most useful when you don't exactly know what the final answer should be; that is, you're trying to find a solution to a problem, but you don't have a specific goal-state to achieve.

It is more of an exploratory technique in which you grope your way through unknown territory to find an answer. Backward chaining, in contrast, is most useful when you know exactly what your goal-state should be. It can dramatically reduce the effort involved by pruning out all solutions that do not end up with the correct answer. Backward chaining is nearly always appropriate for path-planning systems because in planning a path you almost always know the destination of the path. In the example above, it eliminates any path that includes a plane ride to Seattle, for example, since Seattle is not the ultimate destination. In a forward-chaining solution, a plane ride to Seattle would have to be at least considered since it is (more or less) traveling in the right direction.

An intelligent system doesn't have to choose between only these two reasoning techniques, however. More complex systems can reason in both directions at once. The idea here is to construct paths that depart from New York, and to simultaneously construct paths that arrive in Los Angeles. When a path leaving New York meets a path that arrives at Los Angeles, a solution has been found.

The travel-to-L.A. scenario sounds simple, but in fact it can be quite a bit more complex than you might think. The additional complexity arises from the unexpressed additional constraints on the solution. For example, if you are a person who dislikes flying, all the paths that include a plane ride must be eliminated. If time is of the essence, paths that involve substantial cross-country driving in an automobile or bus would similarly be removed from consideration. Constraints allow the system to perform some heuristic "pruning." The term comes from considering the possible solution paths as the branches of a tree; lopping off a few branches here and there makes for substantially reduced search requirements.

Heuristics frequently can be applied directly to solving a problem. A heuristic is a general rule of thumb that provides guidance on finding a solution. Heuristics need not be correct all the time; they just need to improve the likelihood of finding the correct solution. In the New York to Los Angeles problem, for example, you might apply the heuristic that if the distance to be traveled is greater than 500 miles, you should try to travel by train or plane; if greater than 1000 miles, you should try to travel by plane. This heuristic is an extremely powerful one for this problem because it immediately determines that the majority of the distance between New York and Los Angeles should be expected to be traveled by plane. That dramatically reduces the possible paths and concentrates the effort expended into discovering how to travel from your start position to the New York airport, and how to travel from the Los Angeles airport to your final destination.

Heuristics don't necessarily have to be completely built-in to a system when it is initially constructed. Logic systems can learn rules, just as neural networks can learn patterns. Typically, learning in a knowledge base occurs using a supervised scheme in which the system receives feedback about its performance. The feedback can be from a specific tutor, or the learning procedure can modify itself by watching the environment and inferring its performance level.

Learning heuristics can be based on two kinds of inputs. If the system is provided with a set of examples of correct solutions to a given problem, the system creates a generalization heuristic. In this case the characteristics of the examples are merged and broadened to create a general set of characteristics for correct solutions; the knowledge system pays attention to the similarities among the examples in this case. If, on the other hand, the system is provided with a collection of almost-correct solutions, the system specializes and focuses in to the specific set of characteristics for the solution. It learns that specific features are or are not acceptable in the solution; it pays attention to the differences among examples. Just as with neural network learning, AI learning proceeds in small increments rather than in large leaps.

When a system is learning, it needs feedback on its performance. The feedback gives an error estimate, which is similar in concept—if not in specifics—to the errors used in training neural networks. In a machine-learning program, however, not all errors are treated equally. Most recent AI programs try to assess the import of each error and categorize them by their relative impact on achieving the desired goal. If the error is minor, easily correctable, or does not affect the overall achievement of the eventual goal state, it may be classified as an "ignorable" error. Of the remaining errors, some are "recoverable" in the sense that new actions taken by the system may still allow the network to achieve the goal. The remaining errors are considered "irrecoverable," meaning that they are so severe that the goal cannot be achieved. For example, an ignorable error in traveling to Los Angeles might be that you go to the Delta Airlines ticket counter rather than the American Airlines counter. As long as both companies have flights to Los Angeles, no harm is done and you can purchase a ticket and be on your way. A recoverable error occurs if you go to the Qantas ticket counter at the airport and you find that Qantas has no flights to Los Angeles from New York. In this case you have to "undo" something you did and go to another airline's counter to proceed. Finally, an irrecoverable error occurs if you lose all your money and credit cards on the way to the airport; you cannot get the money back and thus cannot proceed at all.

Yet another problem solving technique can be used for general problem solving. This is "means-ends" analysis, and was developed by Allen Newell and Herbert A. Simon in the early 1970s. Means-ends analysis is a process that involves the following steps.

1. Consider the differences between the current state of the world and the desired state of the world, or goal-state. If the current state matches the goal-state, stop; the problem is solved.
2. Select an operator that has the effect of reducing at least one of the differences found in step 1.
3. If the operator can be applied—that is, if all the conditions needed to apply it are satisfied—use the operator. If not, generate a new secondary goal of achieving the conditions necessary to be able to apply this operator. Use this same means-ends analysis procedure to achieve that secondary goal.
4. Repeat these steps until the goal state is achieved.

These steps are repeated in order until a stopping condition is met and the goal achieved. The name, "means-ends" analysis, comes from the fact that the system repeatedly analyzes the goals (ends) appropriate at the current time, and tries to find a way (means) to achieve those goals.

If means-ends analysis is applied to the problem of traveling from New York to Los Angeles, the system might use the following reasoning:

GOAL:	Travel to Los Angeles (distance > 1000 miles)
MEANS:	Travel by plane because distance > 1000 miles.
PROBLEM:	Cannot apply "travel by plane" because condition "at airport" not met.
SUBGOAL:	Travel to airport so "travel by plane" operator can be applied
MEANS:	Travel by cab to airport
STATE CHANGE:	Now at airport, subgoal satisfied
STATE CHANGE:	Apply "travel by plane" operator . . .

Means-ends analysis is an example of a recursive operation. Recursive functions are those that call themselves when they are executed. In this case, step 3 of the general algorithm specifies that a subgoal be generated and the same means-ends analysis procedure be used to solve that subgoal before continuing. You might think that this process is in danger of becoming an endless loop that never terminates, and in certain circumstances you would be correct. Let's see what happens when a recursive function is actually executed.

Figure 8.1 presents a simplified execution history for a recursive means-ends system. The initial goal state is G. However, G cannot be directly solved because condition A is not met. Therefore, the system produces a subgoal of achieving condition A. When means-ends analysis is applied to subgoal A, however, a new condition, B, must be achieved before A can be solved. This spawns a sub-subgoal B, and again means-ends analysis is applied. Sub-subgoal B has two conditions that must be met before it can be solved, however, and sub-sub-subgoals C and D are generated. Means-ends analysis is applied to each of these in turn and, happily, solutions for both are found. Once C and D are achieved, B can be solved. Once B can be solved, A is solved. Finally, once A is solved, the original goal G can at last be solved and the goal achieved.

Recursive functions frequently appear to make matters worse rather than better when they first begin to execute because they tend to spawn larger numbers of goals and subgoals without satisfying any of them. Once the bottom of the recursive stack is located however, the goals are rapidly achieved and matters improve substantially.

Recursive functions operate by using a stack. A stack in computer parlance refers to a situation in which the most recently generated goal or function is resolved first. It is a "last in-first out" situation; a

Evaluate distance to goal, G.
Select an operator.
Designate a subgoal A to solve.
    Evaluate distance to subgoal A.
    Select an operator.
    Designate sub-subgoal B to resolve.
        Evaluate distance to sub-subgoal B.
        Select an operator.
        Designate sub-sub-subgoal C to resolve.
            Evaluate distance to sub-sub-subgoal C.
            Select an operator.
            Apply operator and solve for sub-sub-subgoal C.
        Designate sub-sub-subgoal D to resolve.
            Evaluate distance to sub-sub-subgoal D.
            Select an operator.
            Apply operator and solve for sub-sub-subgoal D.
        Apply operator and solve for sub-subgoal B.
    Apply operator and solve for subgoal A.
Apply operator and solve for goal G.

Figure 8.1 An example of the execution history of a recursive form of means-ends analysis.

good real-world analogy is a stack of plates in a cabinet. If clean plates are always placed on the top of the stack, they are also the first ones used at the next meal. To use the bottom plate on the stack all the other clean plates have to be removed first; thus, the first plate placed on the stack—the botton one—is always the last plate used. In the case of the example here, the "last in" are sub-sub-subgoals C and D; they are thus solved first. The last goal to be resolved is G, the original one placed on the stack.*

An expert system or other problem-solving program may use any or all of these techniques to solve problems. All these approaches are really search techniques, however, and search has some limitations in its general usefulness. The most appropriate uses for a search strategy occur when the problem is very well defined, with initial states and goal-states, along with any applicable constraints, that are well understood. Additionally, search really can be applied only when there is a known collection of operations the system can perform that have predictable outcomes. For example, the operation "take a plane to Los Angeles" is well defined, and has a predictable result: You change your location to the Los Angeles airport.

Not all problems have such neat and tidy characteristics, however, and search strategies falter badly under such circumstances. For example, suppose the problem is to write a poem. The current state is easy to define: The paper is blank. But what should the goal-state be? The goal can't be defined until the problem is solved, so there is no way, for example, to measure the "distance" between the initial state and the goal state. In fact, search strategies do not work well with any problem that can be termed "creative." While search may do very well to assist in the final production of a creation, it is inadequate to generate the creative impulse to begin with.

Even when the goal-state is well defined, if the system does not know or does not have the operations that may be most useful at solving the problem, the system is likely to fail. For example, suppose in the New York to Los Angeles problem the only operation that the knowledge base knows about is to hop on a pogo stick. Clearly the resulting path solution is going to be both nonsensical and impractical.

*Recursion is often used in AI programs which is why Lisp is such a popular programming language for AI. Lisp is one of the first computer languages developed, and recursive functions are extremely easy to implement in it. The language is not used much in more conventional applications, however. Traditional-language programmers frequently comment that the name "Lisp" stands for "Lots of Irritating and Silly Parentheses" because of the long, nested sets of parentheses characteristic of Lisp programs.

On the other hand, search strategies provide the very heart and soul of many a successful, practical application; they are frequently used in the design of expert systems, for example. These AI programs, which reason through collections of rules and facts to draw expert-quality conclusions, have shown impressive performances across a very broad range of problems. It is certain that an android will make lavish use of this technology to assist it in making its way in the world.

If search strategies are insufficient to solve the real-world problems that an intelligent android will face, what can be done to supplement them? One obvious possibility is to use logic to reason through the available data and develop a solution. This technique is very appealing to a programmer because computers are very good at logic, and it seems that this is an ability that can be exploited in developing an intelligent computer system.

For a change, the obvious answer is a very good one. Many AI programs deal with logical reasoning, and they perform these operations far more efficiently than people. Logic is not a single solution, however; there are dozens of different kinds of logic. Let's consider just a couple of the more common varieties for the moment.

The most commonly used logic is predicate calculus, a system of formal logic that is widely employed in AI applications. Probably the easiest way to describe it is to consider a specific example. A brief knowledge base might contain the following objects and facts:

Delilah is a cat.
Mittens is a cat.
Delilah is black.
Mittens is gray.
All cats are furry.

We might query this knowledge base by asking, "Is Delilah a cat?" The knowledge base responds "true" to this query.

To be useful, predicate calculus must be able to do more than simply retrieve information we already know. Various kinds of queries can be submitted to the knowledge base. For example, a query can ask if an object exists that matches a particular property, such as, "Is there an object, such that the object is a cat?" The knowledge base replies "Delilah and Mittens" to this query because those objects have the specified property. Properties can be combined in queries using any of several logical operators, or even combinations of operators. These operators include AND, OR, XOR, NOT, NOR, NAND, IMPLIES, and a number of others. All these operators except NOT require at least two arguments or conditions that the system tests; NOT is a unary operator, meaning that it accepts only a single argument.

Consider a few simple examples of statements based on the knowledge base above:

(Mittens AND Delilah) is-a (cat) —> true
(Mittens AND Delilah) is (gray) —> false
(Delilah OR Mittens) is (black) —> true
(Delilah OR Mittens) is-a (cat) —> true
(Delilah XOR Mittens) is-a (cat) —> false

The first statement is "true" because both Mittens and Delilah are indeed cats. The second is "false" because both cats are not gray; only Mittens is gray, while Delilah is black. The phrase (Delilah OR Mittens) returns "true" if either or both are black; for similar reasons the fourth statement also returns "true." The last example (Delilah XOR Mittens), returns "false" because XOR is the "exclusive or" function. It returns true if either of its arguments is true, but *not* if both are true; since both Delilah and Mittens are cats, the XOR function gives a "false" response.

From these simple building blocks much more complex relationships can be expressed. For example, the statement "all cats are furry" can be expressed more logically as

For all *x* where (*x* is-a "cat"), then (*x* is "furry")

With the statement rephrased in this manner it is easy to see how logical reasoning can occur. If we want to test the notion, "Is Mittens furry?" we can recast the query as

"Mittens" is "furry"?

The question mark indicates that this is a statement whose truth or falsity is to be tested. The only rule that exists that has anything to do with concluding that something is furry is the statement:

For all *x* where (*x* is-a "cat") then (*x* is "furry")

To make this match the query, the *x* in the rule must be identified with "Mittens." This process is called "binding" the specific object, Mittens, with a temporary variable, *x*. To prove the query we now only need to satisfy the if-clause of the statement, in other words, we have to demonstrate that "Mittens" is-a "cat." Happily, the knowledge base yields this statement as one of the facts that is known about Mittens. Thus, we can confidently assure ourselves that Mittens is indeed furry.

This example is exceedingly simple. Real life problems are not nearly so easy to solve. One of the characteristics that makes realistic problems so hard is that the information in the knowledge base is rarely so cut and dried. Facts are rarely known or presented in black-and-white, true-or-false terms. Instead, we are more likely to know that when the sky is dark and overcast it is *probably* going to rain.

Systems that have knowledge bases must always keep track of the degree to which each "fact" can be relied on. For example, if a knowledge base holds the "fact" that it is going to rain today, the degree of confidence in that fact needs to be specified. Generally, the confidence or certainty factor for a fact is estimated initially at the time the information is added to the knowledge base. In the case of the day's weather, the certainty factor might be 60 percent, because the local newspaper's weather forecast specified a 60-percent chance of rain. Confidence levels are not fixed values, however, and as new information is added to the knowledge base, sometimes it affects the certainty with which other facts are held. If you get a telephone call from your friend across town who tells you that it's just starting to sprinkle outside, you might increase the certainty factor that it will rain where you are to 90 percent. And if you look outside and see water falling from the sky, the certainty factor changes again to 100 percent.

One of the most vexing problems in dealing with knowledge-based systems, in fact, is how to deal with certainty factors, and how to change them consistently and reliably. Typically, a certainty factor is expressed as a probability, which means that it varies from 0 (the fact is certainly not true) to 1.0 (the fact is certainly true). A probability is an estimate of the likelihood of something occurring. The classic example is the flip of a coin. If the coin used is an honest one (not weighted or in any other way biased), the coin has equal probabilities of coming up heads or tails. This is expressed by saying that the probability of getting heads is 50 percent (0.50); similarly, tails also has a 50 percent probability. The two probabilities must add up to 100 percent since these two possibilities account for all the possible ways the coin might land when we flip it.*

Probabilities are computed by accounting for all possible outcomes and then counting the number of outcomes of a particular type. For any given event, the total probability across all possible event outcomes must be 1.0, or 100 percent, just as the coin-flip

*In this example and the following ones, I ignore the possibility of the coin's landing on its edge. This is not unfair, since such an event almost never happens in real life. Even if it does, we can easily force the coin into a heads-or-tails result by blindly (i.e., without looking at the coin) knocking it over onto its side.

probabilities added to 1.0. If the probabilities do not add to exactly 1.0, either some event outcome has been omitted, or the estimates of the probabilities of the outcomes are inaccurate.

Suppose you flip two coins either simultaneously or one at a time. What is the probability that both come up heads? The possible outcomes for these two events are shown in Figure 8.2. Four outcomes might occur, and in only one do both coins come up heads. Thus, there is one chance out of four that the result is two heads.

Well-known principles exist for combining and merging probabilities in an organized fashion, but they all assume that probabilities can be estimated reasonably accurately. In the case of the two-coin flip, because the outcome of the first coin has no effect whatsoever on the outcome of the second coin, the two coin flips are independent events. As a result, their individual probabilities for resulting in heads multiply together to generate the probability that both are heads. Thus, 0.50 times 0.50 is 0.25, which is the probability of both coins being heads (or both being tails).

The difficultly with using simple probabilities for knowledge-based systems is that people seem to find probability theory counter-intuitive. For example, if you have flipped ten coins in a row, each having a heads outcome, which result is most probable in the next flip, heads or tails? Many, if not most, people assume that tails is now "due" and guess that tails is more likely than heads. The fact is, however, that on the eleventh flip, just as on the eleven-hundredth flip, heads and tails have equal, 50 percent probabilities of occurring. The fact that you have experienced a long run of heads has absolutely nothing to do with the outcome of the next flip—if the coin is an honest one.

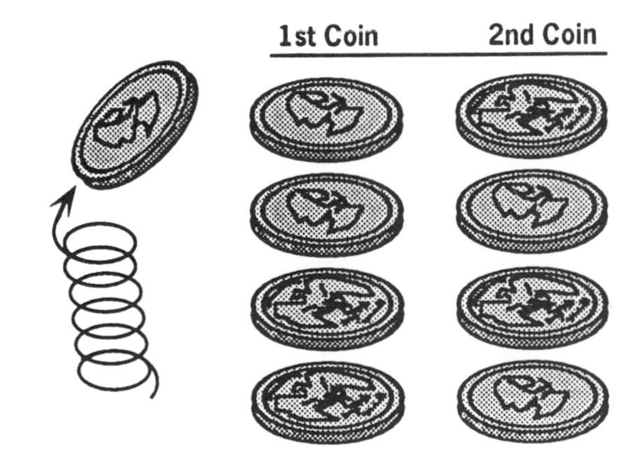

Figure 8.2 The possible outcomes of the event of flipping two coins.

What this means for knowledge systems is that it can be extraordinarily difficult to get human experts to assign a probability to a fact. Furthermore, experts also usually have trouble determining in a self-consistent fashion how much a certainty factor should change in light of new information. Most people simply have a lot of trouble thinking in terms of probabilities. This means that it can be difficult to construct a knowledge-based system using probabilities as certainty factors.

A number of alternatives to standard probability theory exist for use in dealing with certainty factors. One simple approach is the notion of fuzzy values.

Fuzzy values have the intriguing ability to deal with the sloppiness people use in talking about real-world objects and events. While people rarely go around saying, "I think the probability that I'm going to get that job offer is 83 percent," they do say something more like, "I have an excellent chance of getting that job." What does "excellent" mean in this case? Eighty percent? Seventy-three percent? Ten percent? Most people would assume that 10 percent likelihood does not qualify as "excellent," but anything greater than 50 percent might be termed excellent, depending on the person doing the interpretation. And if the speaker is someone known for exaggerating, even 10 percent might not be a bad interpretation.

Fuzzy systems don't have to go through this kind of agonizing over exact numerical values. Instead, English-language terminology can be directly interpreted as fuzzy categories. Fuzzy categories such as "poor," "good," "fair," and "excellent" are used, but the specific numerical values assigned to each category are much less important than the relative order in which each ranks. Interestingly, very rarely does a person have trouble ranking items from worst to best, or most to least; it is only the task of estimating a specific probability that causes problems. Reasoning using fuzzy categories takes advantage of this ability to determine relative order easily, without demanding numerical estimates from human experts.*

Fuzzy certainty factors are far from the only technique used to overcome the probability barrier. Researchers have developed a wide range of other methods that effectively deal with this situation. Such efforts, called plausible reasoning, or reasoning with uncertainty,

*Interestingly, one of the reasons human experts frequently have trouble with probability estimates appears to be an unwillingness to commit to specific numbers, with which their own performance can be measured. By using fuzzy categories, the "number barrier" is removed and experts often become much more comfortable with sharing their expertise.

have enabled systems to cope with the vague, ill-defined terminology we all use every day.*

All these efforts are effective in generating problem-solving applications. One of the earliest AI programs, the General Problem Solver, was able to do mathematical proofs of theorems. Heuristic search strategies are quite effective whenever the problem is well defined, with known operators and defined goals and states. Logic can be used to reason through a solution based on information maintained in a knowledge base. Probabilistic and fuzzy reasoning processes enable programs to deal with uncertainty in the information and rules stored in knowledge bases. These techniques work very well, but they don't seem to provide a complete solution to the general task of solving problems.

There are times when people use none of these methods to solve a problem. Probably everyone has had the experience of puzzling at a problem for some time, apparently making little or no progress with it, and then suddenly "seeing" the solution. It is as if the answer pops out of nowhere, fully formed in the mind. This experience, sometimes called the "aha!" experience, appears universal in humans. Finding the solution seems somehow out of our control, as if it is being developed somewhere in the subconscious. Furthermore, rarely is this experienced with an easy or straightforward problem; the aha! experience seems linked mostly with difficult or unusual problems. All conscious musings about the problem seem to have little relevance, then suddenly, as if a bright light comes on, the answer pops to the top of the mind. Is it possible to build a computer program that can experience the aha! phenomenon?

No one yet knows the answer, largely because no one really knows how it works in people. Some specific details of how it may occur have been discovered in research with people, however. One key characteristic that the aha! experience seems to have is the notion of analogies. Usually people who experience it draw analogues from the unfamiliar problem they are faced with to something else that is easier or more familiar to them. If the analogy is a reasonable one, it may permit the person to use a solution appropriate to the analogue on the problem.

Humans are not the only animals who can solve problems with analogies; the chimpanzee mentioned at the beginning of this chapter clearly drew an analogy between the pole and a tree. In Chapter 11 I

*For a superb discussion of a variety of such systems, see Judea Pearl's *Probabilistic Reasoning in Intelligent Systems,* listed in the Suggested Reading list at the end of this book.

consider this skill more closely in the discussion of creativity in an android. The ability to draw an analogy between an unfamiliar situation and one or more familiar situations has already been provided in some AI story understanding programs, so at least this skill can be assumed in an intelligent android.

The aha! experience is not well understood in humans, but more is likely to be involved in it than a mere ability to draw analogies to familiar objects and concepts. People who experience it nearly always speak of the experience as they would a sensory one: They suddenly "see" the problem in a new light, for example. Furthermore, it seems to be an experience that is somehow outside the normal cognitive centers. Research has discovered that people who are working on a difficult problem rarely, if ever, know that they are about to experience an aha! breakthrough until it happens. Even up to the last few seconds before the answer is "discovered," people report feeling that they are far from a solution. This is in strong contrast to problem-solving experiences that do not involve an aha! experience. In such situations people report that they are making steady progress on the problem, and can predict reasonably accurately when they can be expected to provide the solution.

Such research reports are not easily explainable, but are intriguing nevertheless. The inability to explain how an answer was reached, and the lack of awareness of the progress that is being made on the problem at a conscious level smack strongly of the hallmarks of a solution being generated by a neural network rather than anything equivalent to AI symbolic computation. Neural networks, unlike AI programs, are very poor at "explaining" how an answer is reached. In a knowledge-based program, the exact chain of reasoning can always be extracted so that the specific steps taken to solve a problem are always explainable. In contrast, however, a neural network doesn't reason its way to a solution at all; thus there simply is no sequence of steps to report on. A neural network locates an answer because it locates an answer; any other explanation must be confabulated more or less after the fact. And since neural networks do not reason, there is no way to report progress on a problem until the answer is actually found.

Furthermore, the fact that people consistently report that finding the solution to a problem is like a sensory experience rather than a cognitive one gives further evidence that the experience may be based not in the logical part of the brain, but in the sensory portion. Many scientists believe that even if the "higher" parts of the brain perform symbolic computation as in an AI program, low-level sensory processing is performed largely by nonrational neural networks that have simply learned how to respond to appropriate external stimuli.

Based on these notions, it is possible to speculate that a difficult problem may somehow get "turned over" to the sensory system to work on when there is insufficient information to allow the more cognitive centers to solve it. This might also help explain why such problems are so frequently solved overnight when we sleep. During sleep the sensory systems may have less to do than during an active daytime, and thus be able to devote more resources to solving the problem. My speculations on this subject are merely that and nothing more at this time, but to me they do seem to explain many of the characteristics of the aha! experience.

If this explanation, or something similar to it, proves with further research to be valid, there should be no reason why an android could not also have an aha! experience. It should be possible to provide it with a control system that allows very hard problems to be shifted over to its neural network sensory processors for solution. The android may then also have to be provided with "down time" equivalent to human sleep so that the sensory system could work on such problems without endangering the android by depriving it of sensory processing capability.

An intelligent android can be expected to have very good problem-solving capabilities, although not as good as those of a person initially. AI has decades of experience in developing systems that can solve a variety of problems using search, reasoning, and knowledge-based systems. While those capabilities are not likely to be able to solve every problem facing the android, they certainly ought to be able to find answers to most everyday problems. And if my speculations on the aha! experience prove to have some merit, an android may be just as capable as you or I of waking up in the morning and spontaneously just "seeing" the answer to a difficult problem.

If this is the case, we'll need to make sure the android has the word "eureka" in its vocabulary.

$\geqq$ **9** $\leqq$

Speaking in Tongues

Talk is cheap because supply exceeds demand.
Unknown

Of all the fine arts known to mankind, one of the finest is the art of conversation. People feel compelled to exchange their thoughts and feelings through discussions, conferences, interviews, debates, and plain old fashioned arguments. Our skill with language is one of the blessings—and curses—of our kind, and any device that seeks to converse with us must be able to share our language. And since the majority of human communication occurs verbally, the android must speak as well as compose text messages. It must talk to us, understand what we say to it, and comprehend the meaning behind the words it perceives. In essence, it must understand speech.

The problem of understanding speech is actually three problems in one. First is the problem of speech generation: that is, the android must be able to produce the appropriate sound waves that correspond to a message it needs to convey to a person. Second is the problem of understanding verbal speech. A conversation is (usually) not a one-way street, and for effective communication the android must be able to understand what is said to it. This capability has a number of pitfalls that must be overcome, including problems in accent, pitch, and timing. Finally, having a meaningful conversation is more than exchanging similar sounds. A conversation requires that the meaning behind the words be understood and comprehended by both parties. If we want the android to be able to talk, it must be able to understand not just the words, but what they mean and imply. It must understand not just speech, but language.

Consider the problem of making a system that can generate human speech—let's say English speech to be specific. This is not nearly as easy to do as one might suppose. Early commercial speech generators frequently sounded tinny, wooden, and unexpressive, with odd

Figure 9.1 2001*'s HAL computer had sophisticated speech understanding capabilities. He was even able to read the lips of astronauts David Bowman and Frank Poole as they discussed their concerns about his performance. (Photo supplied courtesy of Turner Entertainment Company.*

tonality and pitch patterns. Anyone who has ever called directory assistance can vouch for the odd tonality and pitch patterns they generate. Furthermore, the behind-the-scenes preprocessing that must go on before they can work properly is often formidable.

A typical commercial speech generator works by assigning fixed sound patterns to letter combinations. It follows a set of rules in determining which sound pattern applies to each letter (or letter combination, such as "th"). The stresses within a word may be either ignored or, when imitated, often generate inappropriately accented syllables. For some languages this simple approach might not be too bad. Spanish, for example, has extremely uniform spelling and pronunciation of words. Each letter has only a few possible ways to pronounce it, and those pronunciations are determined by strict rules that are easily encoded in a speech generation system. But for a system that is trying to pronounce English words, the problems are manifold.

English has the unfortunate characteristic of having many words that look similar but are pronounced quite differently. "Though," "through," "cough," and "bough" are obvious examples. Even worse, a number of words have identical spellings and quite different pronunciations: "row," a tier of seats in a stadium, and "row," an argument or uproar, look identical but the first rhymes with "toe" and the second rhymes with "how." Similarly "read" can be pronounced "reed" if it means the present tense ("I read a book now"), or it can be pronounced "red" if it means the past tense ("I read a book yesterday"). English is filled with many such anomalous pronunciations through which speech generation systems must wade and must learn. While many good speech generators can pronounce the basic sounds of English words (the "phonemes") correctly, the problem lies in determining which phonemes are appropriate for each word. There are nearly as many exceptions to the rules as there are rules themselves, so the problem needs a non-rule-based solution to supplement a good set of rules for pronunciation.

Happily, such a solution may not be too hard to generate. In 1984, Terrence Sejnowski and Charles Rosenberg developed a system at Johns Hopkins University called NETtalk. While not a complete speech generation system, it demonstrated that such systems are feasible.

NETtalk was an attempt to demonstrate neural network speech generation system capabilities. It was designed to convert the text of a message into the phonemes necessary for a commercial speech generator to do a good job pronouncing the text. For the experiment, a tape recording of a 6 year old reading a paragraph that described a visit to his grandfather was used as the master training data. Sejnowski and his co-workers analyzed the child's recording and determined the correct phoneme codes for each letter in the passage. From the point of view of the neural network, there were 29 possible letter codes: one for each of the 26 letters of English, plus codes for each of a space (a word or sentence separator), a comma (indicating a pause), and a period (indicating a hard stop).

The network was trained by giving it the current letter in the passage and the correct phoneme code for that letter. Because so much of English pronunciation of a letter depends on its context, the network was also provided with the codes for the three letters immediately before and after the current letter. Essentially, the network looked at a moving window that was seven letters wide, and in which the current letter to be pronounced was the central letter. In Figure 9.2, the current letter is the "s" of "is," preceded by "s_i" and followed by "_my".

The network's output was a numerical code that told the commercial speech generator which phoneme to pronounce next. The speech

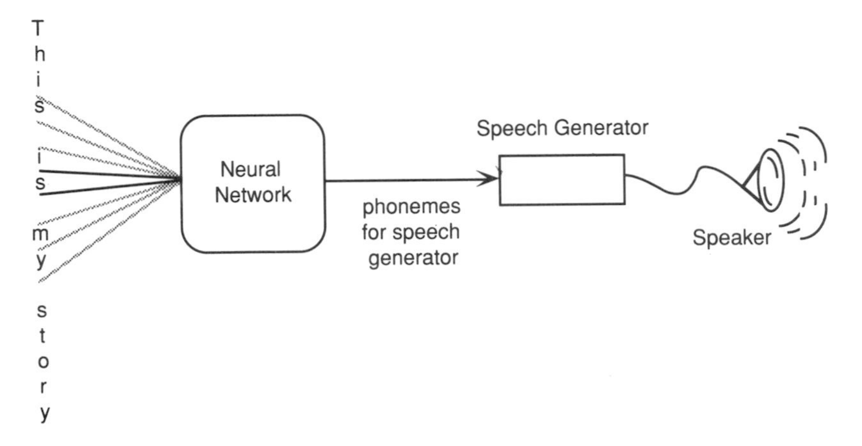

Figure 9.2 NETtalk looked at the text of a story and generated the correct phoneme for a commercial speech generation system.

generator then output the specified sound over a speaker; this output was tape recorded. As with most neural networks, this system learned through repetition. During training, it was given the same paragraph to say, over and over again, with a tape recorder capturing its progress—and this progress was remarkable.

When the system was first started, the sounds it made were completely unintelligible. It sounded much like a continuous drone or babble, with no meaning at all. This was, of course, to be expected since neural networks traditionally begin their training with random connections between their neurodes; if any outputs are correct under these circumstances, it is strictly by chance. As training proceeded, however, the network soon learned to put pauses between words and distinguish between a vowel and a consonant. However, since it initially used only a single vowel and a single consonant, its output had an eerie similarity to the sound of the baby-talk babble of an infant—while it made no sense at all, there was a feeling of hearing words in some strange pseudolanguage. With further training, the network occasionally got a syllable correct, and then an occasional word made sense. As the network refined its understanding of how to speak, it demonstrated greater and greater success, so that it finally spoke as well as the 6 year old who was its teacher.

And how long did it take the network to go from continuous babble to sophisticated pronunciation? The network was fully trained overnight. In a few hours, it literally moved from an infant's jabber to the competence of a 6-year-old child. With current commercial neural network chip hardware, this experiment now can be accomplished with a training time of only a few seconds instead of a few hours, making NETtalk—or an extended version—practical for an android speech system.

When training was complete, the network was not perfect in its pronunciation, just as a child is not always perfect. Its accuracy was about 92 percent, meaning that it correctly pronounced about 92 percent of the words in the training passage. When tested on a passage that was not used during training, it offered the correct pronunciation more than three-quarters of the time. A typical error was to confuse the pronunciation of the hard "th" sound in "them" with the soft "th" sound in "throw"—quite an acceptable mistake in a system only a few hours old.

Sejnowski then performed some experiments on this trained network. He was interested in how successfully the network could handle damage to its weights and interconnections. This is an important issue, because in people, hundreds or thousands of neurons die daily in the brain. Since neurons generally do not reproduce, the total number of brain cells we have available decreases steadily from birth to death. If the circuitry we have has significantly impaired performance when a few cells die or a few connections break, then it might be expected that our mental powers would decrease considerably as we age. Sejnowski wanted to determine how much damage a trained network could sustain and still maintain acceptable performance levels.

The answer was quite astonishing. Sejnowski damaged the network by permitting random changes of varying magnitudes in the strengths of the connections between the neurodes. Remember that information in a neural network is stored not in the neurodes themselves, but rather in the pattern of connections between the neurodes, so damaging the weights was equivalent to dropping memory locations in a computer, or to garbling the memories themselves. Sejnowski found that the system's performance was hardly affected at all until the damage to the weights was in the range of plus or minus half the average connection strength value. And when the damage was of approximately the same magnitude as the average weight, the network was still achieving over 80 percent accuracy in its pronunciation.

NETtalk is certainly not the ideal speech generation system, but it does illustrate the power of the neural network approach to the problem, particularly when combined with the well-developed technology of current speech generators. Building and training the neural network to control the speech generator took considerably less time and effort than building a complex rule-based system would have required. The big problem with NETtalk as a self-learning system was that its inputs had to be carefully pre-digested. While it took essentially the raw text (in an easy-to-encode form) as its basic input, it was trained using phoneme codes that had been translated by hand from the original tape recording. In other words, NETtalk did not figure out for itself how to pronounce the words in the text; it was explicitly told what phonemes went with each character. This is not exactly a horrendous limitation, but it also does not imply any grand intelligence on the part of the system. Still, NETtalk proved that it is possible to create systems that generate natural-sounding speech with technology that is available today. And since the overall NETtalk system was rather small—only about 350 neurodes in the entire neural network—it is easy to conceive how a more complex version would make a dandy speech generation system for an android.

Today commercial speech generators are continuously improving so that the newer systems can do NETtalk's task—convert English text into speech—very well indeed. They are not yet perfect in pitch and tonality, but appliances, vending machines, and electronic equipment that talk to (and back) to us are becoming more common every day.

Speech generation is one skill, but it is much more difficult to do the second language task, speech recognition. This is the process of correctly translating sound waves into a textual message. In speech generation, the conversion from a known text message to a set of sound waves that correspond to that message is reasonably straightforward, as we saw above. But the reverse process is to go from sound to an unknown message, and there are many problems along the way.

Sound is a messy medium in which to work. People speaking the same sentence use different accents and different stresses on the words. The sonogram, or picture, of the actual sound patterns created by different people when saying the same words looks remarkably different in each case. Each person has a unique voice, pitch, accent, rate of speech, and set of vocal mannerisms. Worse, the same person may say the same sentence dozens of different ways, depending on mood, state of health, audience, and where the speaking is done. An actor's stage voice, for example, is generally quite different from his or her everyday voice, and many people's voices are modified when heard over the phone or on a tape recording.

Despite this variability, a speech recognition system must be able to cope with all these possibilities. Human beings have a highly reliable speech recognition system themselves, and are rather intolerant of lesser capabilities. Psychologists have concluded that people will not accept a system that has a performance record less than about 98 percent accurate. If the system cannot correctly interpret spoken communications with at least that level of accuracy, the people speaking to it become impatient with the necessity of repeating their words, speaking more slowly than usual, or making other special efforts to make the system understand.

Speech recognizers generally are rated according to three specific categories. First is the number of speakers the system can recognize; a system can be a single-speaker, limited-speaker, or multiple speaker system. A single-speaker speech recognizer is one that is specifically set up to comprehend only the speech of one person. These are useful, for example, when they are implemented as part of a speech interface to a computer or other device. As long as the computer is only expected to cope with the spoken commands of a single person, this works fine. Typically, users of such systems must go through a training period, reciting some (or all) of the system's known vocabulary. Limited speaker systems are really just a minor variation on this, requiring every eventual user to record his or her speech patterns in the system before it can accept commands. Commercial systems today are either single-speaker or limited-speaker systems; but the truly useful systems are those that accept any speaker. These systems, usually called multiple-speaker systems, can interpret nearly any person's voice, whether male or female, old or young, Boston-born or Southern-bred. Just as people from the Midwest can understand a Southerner (usually!), a true multiple-speaker system can handle nearly all speakers.

The second classification of speech recognizers is based on whether they can process continuous speech, or whether they must have words separated in distinct pauses. People don't speak distinctly in their native tongue. Words are slurred together; connectives are contracted, elided, or just plain dropped. And what is written as "Mary had a little lamb" is more often pronounced as "Marihaddalillam." There is no signal to tell the listener what part of the sound pattern received belongs to the word "Mary" and what to the word "lamb." Somehow people manage to sort all this confusion out and correctly interpret the message, but it is more or less amazing that they do. Most speech recognition systems today cannot handle continuous speech. Instead, they require that each word be separated by a distinct pause, so that the message sounds more like "Mary" "had" "a" "little" "lamb." Obviously, this is not acceptable for daily use; people are far too impatient to tolerate having to break up their sentences in this artificial fashion. However, for certain kinds of applications, such as machine interfaces, this can provide a perfectly reasonable choice. And, of course, a separated-speech system is much easier to build than one that accepts continuous speech.

The third way of categorizing speech recognition systems is by the size of their vocabulary. A limited-vocabulary system may understand as few as a dozen words, or as many as a thousand. It has been demonstrated that most people can live much of their daily lives based on a vocabulary of 600 to 1000 words (in English); this is often called the "basic English" vocabulary. So a system that has a total vocabulary about this large may not be as limited as one might otherwise think. While it is considered a limited vocabulary system, it would probably be able to cope with most ordinary human encounters. Systems that have vocabularies that are significantly larger than this—on the order of 10,000 or more words, for example—are generally considered to have an "unlimited" vocabulary.

Obviously, the ultimate speech recognition system is one that combines the best of all three characteristics: It should be multiple-speaker, continuous-speech, and have an unlimited vocabulary. And it should do all this with at least 98 percent accuracy and at real-time speeds. People manage to meet these goals all the time; can an artificial system do as well?

The answer is "not yet, but soon." One of the world's best speech recognition systems was developed by Teuvo Kohonen, the same man who developed the self-organizing feature map discussed in Chapter 7. Kohonen is Finnish and developed the system at the Helsinki University of Technology under contract from a Japanese firm, so it only understands Finnish or Japanese; nevertheless, it is one of the best to date, and similar principles could be used to make an English-language version. He calls it his voice typewriter, because it takes verbal messages and types them out in text.

Kohonen married several key technologies in the development of the voice typewriter. First, he took advantage of modern digital signal processing techniques to convert sound waves into a standard form. This involved a number of computational steps, ranging from using a noise-cancelling microphone to performing fast Fourier transforms on the signal to break down the sound into a series of frequency components. After a bit of manipulation of the sound signals, the result was input into a neural network, which categorized the sound into known syllables and words. Finally, the categories were passed to an expert system that resolved context issues and made the final determination of certain "hard" cases. This final response was then typed out as a translation of what was said.

How good is the voice typewriter? Quite good indeed. When tested under the most difficult circumstances—multiple speakers, continuous speech, and a nearly unlimited vocabulary—it is between 92 and 97 percent accurate. (When coping with multiple speakers, the system uses the first hundred words from each new speaker as training data for the neural network.) When some of the restrictions are relaxed, such as limiting the input to a 1000-word vocabulary, the system's accuracy rises to 96 to 98 percent, nearly at the minimum required performance level for human tolerance of the system. And all the processing involves a mere quarter-second delay from the time the words are spoken to the time the system has identified them.

The voice typewriter is not quite at the performance levels needed by an android; however, it is very, very close. Its accuracy is not quite as high as it needs to be for people to accept it in daily work, and the quarter-second delay is just a bit too long as well. Still, this system is clearly on the right track. One improvement that Kohonen expects to make a big difference is to use more information in the speech signal itself. When the voice typewriter preprocesses the sound waves by taking a Fourier transform of the raw data, it only considers the information contained in the sound frequencies themselves. But just as in other waveforms, sound has phase and timing characteristics as well. If other networks analyzed these other sources of information in parallel with the frequency network, accuracy could almost certainly be increased.

The Kohonen voice typewriter is not the only excellent speech recognition system in the laboratory today. Other researchers are similarly close to developing systems that understand spoken English, and some of them have made excellent progress in the last few years. Current commercial speech-recognition systems that can deal with multiple speakers are also under development; several telephone companies have speech-controlled directory assistance systems under development, for example, and expect to deploy those systems within a few months. While we don't have an acceptable continuous-speech commercial system as this is being written, it can be expected that a very good speech recognition system is not more than a year or two away.

It is one thing to correctly translate spoken words into text, but it is quite different to understand what the words mean. The third problem of communicating with an android is to have a true, natural language capability in the system. This problem is far more difficult than either speech generation or speech recognition.

Traditionally, we communicate with computer systems through a variety of specialized languages. Beginning with the very first programmable computers, the user told the computer what to do by setting ones and zeros into its memory. The sequence of numbers encoded a series of actions—adding or subtracting two numbers, for example—to be performed. These instruction sequences were difficult to comprehend by the programmer, and thus it was both hard to confirm that the program was correct and very easy to introduce errors into it. Rather than communicating through this fundamental code that the machine itself used—and thus its common name "machine code''—programmers wished for a more natural means of communication.

As a result, a new kind of language was developed that allowed programmers to give brief abbreviations to each of the instructions, and to use mnemonic variables instead of absolute computer memory addresses. So machine instructions changed from mysterious sequences such as ''60485, 26028'' to more readable ''LDA X, JMP start,'' each meaning ''load the value stored in *x* into the A-register, then jump to the memory location 'start'.'' These text messages could not be directly executed by the computer, of course, so a new kind of program, called an assembler, was used to assemble the machine instructions that correspond to the text abbreviations. As programmers created larger and more complex programs, however, they found that assembly language had its own limitations and was still quite difficult to write. They wanted a still more English-like language in which to work.

With this need came a large number of special computer languages such as Fortran, Basic, Lisp, and, eventually, Pascal, Prolog, and C. Each of these used a limited vocabulary of English words, and strict grammatical constructs to allow programmers to write computer instructions in a fairly readable form. Like assembly language before them, these also had to be converted into something the computers could read, so programs called ''compilers'' were created. (Interestingly enough, it was originally thought that writing a compiler was an impossible task—until the first one was written. Now it is a typical assignment in an undergraduate student's second or third year of computer science.) Compilers translate the words and symbols of a Fortran (or other language) program into either assembly language—which an assembler then converts to machine code—or directly into the machine code of the computer. The result is then combined with standard library functions to make a file that is directly executable by the computer.

Computer languages are not like the languages we use in everyday speech. They are highly restrictive in form compared to the natural tongues people use, and they are typically ''context-free.'' Context-free means that the meaning of a particular symbol does not depend on the context in which it is found. A plus sign means ''add'' no matter where it is found in a program, and other symbols and tokens have a similar universal meaning. (Before anyone gets carried away by worrying about such fine points as a minus sign meaning both ''subtract'' and ''negative,'' as in a negative number, it should be noted that within the compiler the minus sign (subtract) and the unary minus sign (negative number) are nearly always considered separate and distinct symbols. A preprocessor in the compiler typically scans the program and assigns unique tokens to each symbol, which might be a standard language token for a minus sign or other reserved symbol in the language, or a program-dependent token for a variable or procedure name assigned by the programmer. Thus the variable "count" and a procedure or function "count," if both are permitted in the same program by the language being used, are actually distinctly different tokens. And no computer language permits two different variables of exactly the same name within the same context.)

Human languages, in contrast, are not context free by any stretch of the imagination; they are context sensitive. Words—especially in a vital language like English—have many meanings, and it is left to the listener or reader to interpret them correctly. A phrase like "I ran . . ." can take on many different meanings depending on the remainder of the sentence: "I ran for office"; "I ran three miles yesterday"; "I ran my stockings"; "I ran the company while he was gone"; "I ran into Judy at the store" are just a few of the possible meanings. Add to this the power of metaphors, innuendos, double entendres, puns, implications, and hyperbole, and it really is amazing that anyone understands anything at all.

The problem with such a dizzying array of possible meanings is that just stringing together a collection of dictionary meanings is insufficient to understand the intent behind the meaning. When people hear a speaker, a great deal of information other than the words themselves are used to decipher what the speaker is trying to communicate to us. Such nonverbal clues as body posture, tone of voice, facial expression, gestures, context, audience, historical background of the situation and speaker, and so on are all intermingled in the process of interpreting the sound into a message. Thus, in *Julius Caesar,* Mark Antony's ironic statement, "I come to bury Caesar, not to praise him . . . for Brutus is an honorable man," through repetition and tone comes to mean exactly the opposite of what the words say. And this is far from an isolated instance. Many of our greatest works of literature and poetry fold meaning within meaning, providing texts that have many levels of interpretation. The true test of speech understanding will be when we can show an android a videotape of *Macbeth* or *Romeo and Juliet* and have it understand all—or at least many—of the meanings behind the play.

Several distinct approaches to making a computer understand natural language have been attempted, and they differ quite a bit. These are exemplified by case generative grammars, semantic networks, and conceptual dependency theory.

One common approach taken to understanding language has traditionally been the case generative grammar approach; it harkens back to your sixth grade English teacher who tried to teach you how to diagram sentences. The words in each sentence are examined in a process called parsing, which tries to fit them into a known sentence structure. This is similar to the way children are taught to analyze sentences by identifying subjects and predicates, nouns, verbs, modifiers, prepositional phrases, and so on. In each case, the function of the word or phrase within the sentence is identified. This requires a knowledge of word meanings, parts of speech, and usage. In this process, particular attention is paid to identifying the action referred to by the sentence, the agent of the action—who or what causes the action to occur—and the object of the action—what the action is done to or on.

This approach often turns out to be of only limited usefulness when trying to understand a story. The problem is that while the absolute meaning of the words can be comprehended by such an analysis, the complete meaning behind the words is often more elusive. While some highly successful systems have been built that understand individual sentences, few if any have been able to build this into an ability to understand whole paragraphs or stories.

A second, more generally useful technique in natural language understanding is the semantic network. Semantic networks represent the meaning of a sentence or group of sentences with a graphical construct that consists of nodes and directed arcs. A node corresponds to an object or event in the story. An arc represents one of the basic predicate operators, such as "is a member of," "is," "has," or "is a subset of." The direction on the arc indicates the specific relationship between the nodes it connects; it is essential that the direction be considered in the network. For example, an "is-a" arc that points from node "cat" to node "mammal" indicates that "a cat is-a mammal," which is a true statement. If the arc points from node "mammal" to node "cat," it means that "a mammal is-a cat," a conclusion certainly not true in general. A specific example may make this clearer. Suppose a semantic network is to represent the following simple story:

> Delilah is a cat.
>
> All cats are mammals.
>
> All mammals have fur.

The first sentence means that object-Delilah is an instance of an object-type-cat category. The second sentence means that for any object of type cat, that object is also of type mammal. The third sentence means that for any object of type mammal, that object also has fur. Figure 9.3 illustrates one way these sentences might be repre-

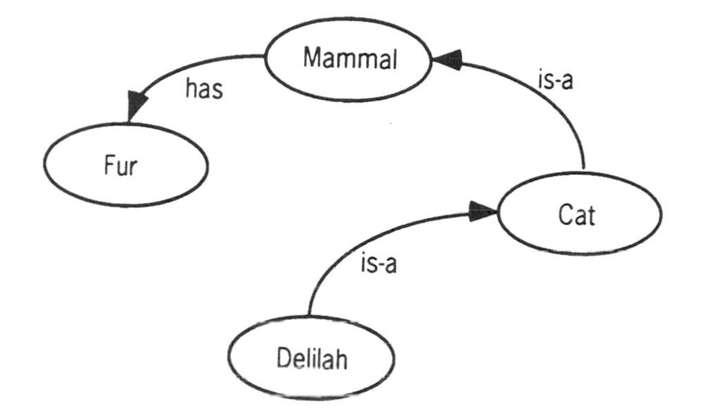

Figure 9.3 A simple semantic network that represents the facts that Delilah is a cat; cats are mammals; and mammals have fur.

sented in a semantic network. In the figure, the objects mentioned in the sentences are represented by nodes in the network. Each arc between the nodes is labeled with the relationship it expresses. For this simple example, the relationships are "is-a," meaning that Delilah "is-a" cat, and that a cat "is-a" mammal; and "has," meaning that mammals have fur.

Semantic networks can be used to draw conclusions about the facts represented directly by the network. For example, one query that might be made is whether Delilah is a mammal. The chain of arcs leading from node Delilah through node cat to node mammal assure us that Delilah is, in fact a mammal. Similarly, we can ask the question, "Does Delilah have fur?" Again, there is a chain of arcs that lead from Delilah all the way over to the "has" arc and down to the fur node, so that the network reveals that Delilah does indeed have fur. Relationships and properties are represented in a semantic network by the fact that an uninterrupted path exists from one side of the query statement (i.e., "Delilah") to the other ("does have fur"). In traversing the network to discover these relationships, however, arcs can only be traveled in the direction indicated by their arrows.

The simple network above can be expanded slightly as shown in Figure 9.4, to represent some additional statements.

Esmerelda is a bird.

All birds are animals.

All mammals are animals.

Birds have feathers.

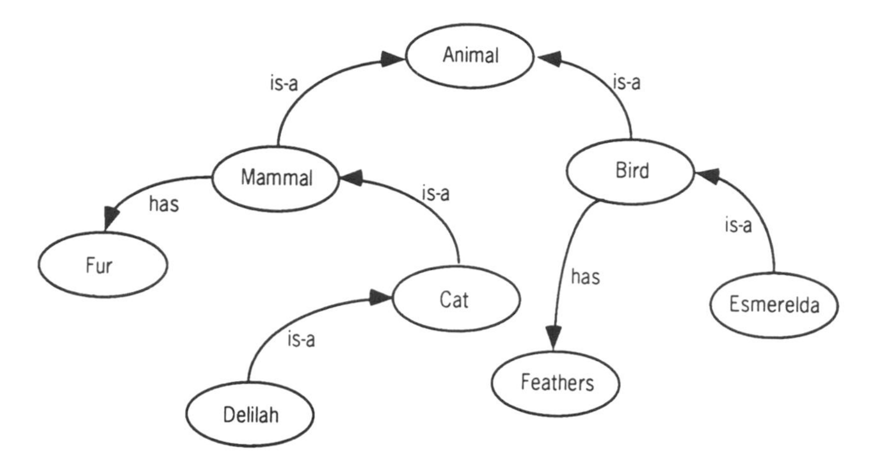

Figure 9.4 The semantic network of Figure 9.3 expanded with the knowledge that a mammal is an animal; a bird is an animal; Esmerelda is a bird; and birds have feathers.

This network can now answer more sophisticated queries, such as "Is Delilah a bird?" "Does Delilah have feathers?" and even "What do Esmerelda and Delilah have in common?" The answers to these queries are found by following the arcs in the network, paying attention to the direction indicated for each arc.

The first question, Is Delilah a bird, can be answered "no" because there is no path that leads from the Delilah-node to the bird-node without violating the direction of at least one arc. Similarly, the second question is also answered "no" for the same reasons. The last question, concerning what Delilah and Esmerelda have in common, is a bit more subtle. To respond to this query, we must follow the arcs from both the Delilah-node and the Esmerelda-node until we find a node where the paths intersect. In the network given, this occurs only at the animal-node. Thus, the response to this query is that both Delilah and Esmerelda are animals.

If the measure of understanding of a collection of sentences is the ability to answer non-trivial questions about what those sentences mean, then semantic networks can be shown to have at least some level of story comprehension. As the networks grow more and more complex, the difficulty of questions they can handle also increases. In the most extreme cases, semantic networks have successfully (and commercially) been used in automatic machine translation systems, translating one human language into another.

Despite these successes, semantic networks have traditionally been plagued by a couple of important drawbacks. First, a semantic network that represents more than a trivial series of relationships very quickly becomes much like a tangled plate of spaghetti, with nodes and arcs twisted into a near-unintelligible mass. Such complex networks can be excruciatingly difficult to support and maintain. Second, a large semantic network tends to take a lot of processing time to answer anything except simple queries. This is because when the response to the query is null (i.e., "there is no such relationship") all possible paths between the start node and the goal node must be searched exhaustively to ensure the accuracy of the result.

More recently, however, with the development of parallel computers and high-speed workstations, semantic networks have become valuable tools in the development of a system that understands the semantics as well as the syntax of language.

There is another technique that works even better for story understanding, and this is called conceptual dependency theory, developed by Roger Schank of Yale University.

Schank believes that almost all the actions we do and write about and talk about can be considered as combinations of only a few basic actions. The idea is that most simple English vocabulary can be described by using only these few actions. The meaning behind a sentence or paragraph thus derives from the concepts expressed by various combinations of these actions, along with their relationships to the agents that perform them—and have them performed upon them. In his view, people ordinarily don't comprehend language by subjecting it to complex sentence parsing operations; instead they analyze it for the concepts expressed by the sentence. This really makes a great deal of sense, because young children can speak and understand language long before they have any idea what a noun or verb might be.

Schank's basic actions include behaviors such as "ingesting," "expelling," "propelling," "speaking," "moving a body part," and "moving an object." People get hungry and eat; they throw a ball or push open a door; they have a conversation, or wave a hand, or place a book on the table. His list of about 20 fundamental actions does an amazingly good job of classifying most of the vocabulary that is considered basic English.

Conceptual dependency really reaches back to the standard journalistic approach of understanding *who* did something, *what* was done, *when* it occurred, *where* it took place, *why* it happened, and *how* it happened. It matters less that each preposition in a sentence can be identified than if the system identifies and understands each of these basic queries, not of the sentence, but of the message behind the sentence. If a system can correctly respond to questions regarding the "five Ws," then it can truly be said to have comprehended the meaning behind the sentence.

Schank and other researchers have devised computer programs that read and appear to understand simple stories using this conceptual dependency approach. For example, a sentence in the text is analyzed to determine the "agent" of an action (the person or thing that performs the action), the "object" of an action (the person or thing the action is performed upon), the "destination" of the action (if it involves movement through space), and of course the specific action itself. Actions themselves are identified by contextual clues. For example, if the sentence has the word "eat" in it, the program might assume that the basic action was one of ingesting something. In this case it then searches the surrounding text for something edible (the object of the action) and a person or animal that does the eating. Other key words might be "restaurant," which could call into play one of the general restaurant scripts,* and thus institute a scan for words in the text that fill various roles in the script. Thus, a sentence like "Bob went to the store" might result in the agent slot identified as "Bob," the action as "move-through-space," the object as "Bob," and the destination as "the store."

Several computer programs built around these ideas have been shown to be quite successful at reading and understanding the text of simple stories. A few have been able to read and understand news stories, either directly from the daily newspaper, or as input from the various wire agencies. (This should not be at all surprising when one considers that journalists are trained specifically to explain the who-what-when-where-why of the news.)

How do the researchers know that the programs truly understand what the stories are about? This is tested the same way we test students on their reading comprehension: The researchers give the program an examination. If the program correctly answers reasonable questions about the subject of the text "read," then it is assumed that it understands the material. The questions in such comprehension tests are designed to do more than elicit the facts detailed in the article; they require that the program draw inferences from the information presented and thus demonstrate more than mere memorization. No computer program has yet been absolutely perfect at this task, but surprisingly good results have been achieved using conceptual dependency and similar approaches.

One problem with this technique of developing story understanding is a lack of consistency. There have been any number of lists of basic actions drawn up by various researchers (and sometimes by the

*Scripts are discussed in detail in Chapter 10; in essence they provide a template for specific situations, giving a guideline for interpretation of the events that occur in that situation.

same researcher at different times). It is not at all clear exactly which actions are basic and which are not. For example, if you walk to the station, should that be considered a "move-through-space" action or a "move-body-parts" action? Or both? Even worse, different people working from the same list of basic actions can analyze the same sentence in two (or more) very different ways. And when the entire context of a story is taken into account, the likelihood of multiple, differing versions of the analysis is only compounded. This lack of firm definition that is ever-present in English, while providing much of the richness of the language, is excruciatingly difficult for researchers to cope with.

There is a famous—or infamous—test of intelligence devised by Alan Turing called the Turing test. The idea is that a person called the examiner is placed in a room with a communications device of some sort—perhaps a keyboard and monitor. In a separate room, out of sight of the person, is either another person or a computer. The examiner types in a message, to which the device—either a person or a computer—responds. A conversation follows between the two parties, with the examiner asking whatever questions he wishes, or providing whatever comments he desires. If the examiner cannot distinguish, through any means that involves only a conversation with the other party, whether he is communicating with a computer or another person, then the system passes the test and is declared "intelligent."

The validity of the Turing test has been hotly debated in the world of computer science for decades. To date, no system has formally passed the test, although a number of surprisingly simple programs do very well at it. A classic example is Eliza, a relatively simple Lisp program that acts as a pseudopsychiatrist. Eliza's "conversation" consists primarily of echoing the user's sentence back as a question, using predefined key words as cues. For example, if the user types in "I dreamed about my mother last night," Eliza might respond with "Do you like your mother?" If no key words are found in whatever the user says, Eliza just rephrases it as a question. "My horse kicked me" might convert to "Why do you think your horse kicked you?" After a very few exchanges of this sort, users almost invariably get an eerie feeling of talking to a little person inside the computer. Yet in spite of this parrot-like echo capability, programmers who look at Eliza's "innards" nearly always concede that nothing at all in the program demonstrates any sort of understanding of the conversation's meaning. And most versions of Eliza can be easily tripped up by using a complex sentence structure or an unusual word order, causing the program to ask a nonsensical question.

A somewhat more sophisticated system called Racter (published by Mindscape) was commercially available for the Macintosh for several years. While not much more sophisticated than Eliza—and still subject to an occasional misinterpretation of the user's responses—Racter could carry on long conversations and even talk out loud using the computer's built-in speech synthesizer. His voice was odd, with a strange atonality, and his statements and stories made a skewed kind of insane sense, but there was definitely a feeling of talking to a personality rather than a machine. It was a bit like talking to a slightly schizophrenic literary genius who made wild leaps and connections, and even occasionally backtracked to previous topics. ("Earlier you said that you liked dreams. Would you like me to tell you a story about my dreams?") Racter had the ability to create stories in his own, somewhat bizarre style, but which generally made sense at some level—at least as much sense, for example, as Lewis Carroll's poem "Jabberwocky." Racter moved from being an "it" to being a definite "he," and acquired an apparent temperament, sense of humor, and disposition as well. It remains to be seen whether such characteristics are universal once we truly have natural language systems.

We soon will have systems that can speak to us in natural tones of voice, that can read text messages with near-perfect pronunciation. We also will soon have systems that can listen to what we say and convert the sounds into the letters and words of a message. With the introduction of new Macintosh models in late 1990 and early 1991, sound and speech have become an intrinsic part of the operation of these personal computers. A commercial device called the Voice Navigator from Articulate Systems, Inc. already allows users of Macintosh computers to talk their way through operation of many software packages. At least from Apple Computer's perspective, speech understanding has arrived at the commercial level.

But while both speech generation and speech understanding are necessary for an android to have good communications skills, they are not sufficient. We must give it language—and not just the verbal skills of a parrot. With language, even though there may be a limited vocabulary or other performance restrictions, an android begins to acquire characteristics that mark it as an individual. Language provides not only communication, but also the possibility of a real personality.

≧ 10 ≦

Sense and Sensibility

The intelligent man finds almost everything ridiculous, the sensible man hardly anything.

Goethe

Could an android be trusted with the simplest of common sense tasks? Will it have the sense to come in out of the rain? Just how many androids will it take to change a light bulb, anyway?

Such questions are not as frivolous as they might seem, because one of the greatest challenges scientists face in building a working android is endowing their creation with the simple sense of a duck—not to mention a person. Today's robots and computers are not well known for their common-sense capabilities. Anyone who has had even a nodding acquaintance with a computer is likely to be all too familiar with the insane lack of rational behavior that they exhibit on occasion. The simplest tasks are occasionally made complex, if not downright impossible, all because the computer lacks the smallest particle of common sense. And this lack occurs not just in the deep recesses of the Social Security system—which occasionally decides someone is deceased, in spite of living evidence to the contrary—or bank and credit companies, which have been known to spend twenty-nine cents in postage and many dollars in employee time to track down a vicious miscreant who underpaid their bill by ten cents. No, it is not just such behemoths that lack common sense; all of today's computers have a basic inability to make reasonable judgments about the world around them.

We have all, on one occasion or another, been bitten by a computer's lack of sense and forcibly drawn into a world gone haywire, from which we can only escape by convincing a real person to help us out by exercising reasonable judgment. In spite of the enormous gains in basic computational power there are literally no major systems in business that are run solely by a computer. Nor is it likely that any important processes will be entirely computer controlled until we find a way of implementing common sense in a machine.

Why is this problem so hard? Some years ago, researchers in AI were convinced that their logical, cognitive approach was certain to result in common-sense computers within ten years. Well, it is now a decade or more later, and no such beast is yet commercially available. Why is this so? How could all those scientists be so wrong?

The fact is that common sense is not nearly as "common" as we would like to believe. In a person, it really consists of all the world knowledge experienced and absorbed by that person over his or her lifetime to date. Furthermore, common sense is not merely composed of facts about the world—like the fact that objects fall to the ground if they are not supported—but also includes a whole realm of beliefs and theories about how the world works and why it works that way. For example, to a person raised on a farm, it is only "common sense" not to approach a mule unexpectedly from the rear—at least if you don't want to be kicked. To a city slicker, there is nothing common at all about such knowledge. On the other hand, a country bumpkin might not have the common sense to keep his wallet in an inside jacket pocket on a crowded subway, something a city resident would be likely to do without even thinking about it, if he carried a wallet at all. Common sense, you see, is not just a matter of objective facts about the world, but is context sensitive—what is common sense in one environment may be just plain dumb in another.

People get a lot of humor out of different varieties of common sense. Movies and TV are constantly using it as the basis for a plot. The popular old TV series "Green Acres" had as its primary premise the uprooting of two urban natives to a backwoods country environment; much of its humor derived from seeing how the city slicker's common sense failed to translate reasonably to a country environment. Just the opposite was the premise for the "Beverly Hillbillies"; here the country bumpkins were displaced to the city, and the humor came from their attempts to deal with city problems using country common sense. More recently, there have been movies like *Funny Farm, Baby Boom* (both with city dwellers displaced to the country) and even *Private Benjamin* (a spoiled rich girl dumped into the harsh environment of Army boot camp), all of which point up the fact that common sense is only sensible within its appropriate environment. In an inappropriate setting, common sense can be hysterically funny—or tragic.

People derive their understanding of the world from many sources. Much of it comes from the training we receive as children. Parents teach their children how to buy a car, a house, and groceries; how to keep house, shop for clothes, and do the laundry; what books to read, television shows to watch, and social causes to espouse. All this is done, either explicitly—"Charlie, don't wash the pots and pans before you do the glasses!"—or by example as parents and children live out their daily lives. So most common sense derives from experience, either the individual's direct experience, observations of the experiences of others, or the lessons taught by parents, teachers, and peers, all of whom pass along their own experiences and beliefs.

What we are really talking about here is an understanding of the world and how it works. An intelligent android must have some reasonable world view to operate successfully outside the laboratory. If we really want these devices to be useful to the general public and in broad situations, then they must be endowed with the common sense appropriate to their world. Obviously, just as the city slicker and the country bumpkin have widely varying notions of what is sensible and what is not, an android designed to work on a farm should have a different world view than one designed to clean a house, which should also differ from one designed to fly a plane. Such specializations are key to making useful androids as opposed to amusing novelties.

An android can learn common sense by experience, just as people do, or can have much of it provided as part of its basic system. For example, we might want to build in a notion of some fundamental laws of physics, so that it would understand the mechnaics of motion and gravity without having to learn them from experience. We could, of course, insist that the android learn about gravity by itself, but that might lead to a fatal case of what is sometimes referred to as the Grand Canyon Effect. This means that we want to pre-teach the android not to step off high ledges, so that it does not try to step off the side of the Grand Canyon—once. Without any pre-knowledge of possible consequences, there would be nothing to keep an android from trying to reach the bottom of the canyon the most efficient way possible—straight down. If we insist that the android learn such details of the way the world works from its own experience, the attrition rate is likely to be nearly 100 percent. (We could mitigate this somewhat by collecting a group of androids and having them watch one of their fellows step off the edge of the cliff and observe the consequences, but that still seems an unnecessarily wasteful—and macabre—means of getting the point across.)

The problem of common sense boils down to a problem in representing knowledge. What we need to do is provide the android with a flexible, reasonably complete database of information about the world and how to cope with it. The information contained in this database must consist of facts, such as the physical and legal laws of the land; conventions, such as the rules for proper social behavior; and beliefs, such as opinions on the relative merits of two kinds of car.

The real problem in providing this database is not so much assembling the information as it is determining how it is to be stored so that the android can access it easily and rapidly in appropriate circumstances.

People store information—memories, in other words—in an astonishingly efficient system. Access to information is nearly instantaneous in most day-to-day situations, and only rarely does a person have to exert a conscious effort to recall a particular item. In fact, such conscious, effortful searches often fail, when an unconscious, casual search succeeds. Nearly everyone has had the experience of trying hard to remember something, only to have the desired memory stay frustratingly out of reach, and yet recall that same memory hours later when doing something entirely different.

In spite of such occasional frustrations, however, most people find their memories to be generally reliable and efficient. Even the elderly, the people most often assumed to have poor memory, have been shown to have excellent recall skills. Their reputations for "losing their memories" with age seems now to be due to illness (such as Alzheimers disease, which affects brain functions in a devastating fashion), to a lack of attention to what is to be remembered, or even to simple lack of practice in memorizing information. While children and young adults often are placed in situations where they must memorize facts and information, this happens much less often with the elderly. As a result, it is possible that much of the memory loss in healthy senior citizens can be overcome by simple practice in recalling information.

While we do not yet fully understand how human memory recall capability works in all its detail, we would certainly like an android to have a similar easy access to important information. Certainly much of the answer must lie in the manner in which information is stored in thc brain.

One possible approach to implementing common sense is by including appropriate expert systems in the android. Expert systems can be viewed as databases that have information encoded as a collection of if-then rules. Such rules provide expert assistance in problem solving or other tasks. The inference engine of the expert system decides which rules are currently pertinent, chooses at least one of those to execute (i.e., perform the "then" clause of the "if this is true, then do something" rule), and otherwise keeps the expert system running. Expert systems have the ability to reason, either by using forward chaining (from the current state forward toward the goal), or by using backward chaining (from the goal back toward the current state). In either case, expert systems can solve problems within their domain of expertise and provide a fine environment for implementing commonsense knowledge.

There is a bit of a problem using expert systems for common sense, however, and that is often called the "mesa effect." This comes from the shape of the performance curve of an expert system. A graph of an expert system's relative performance compared to the domain of knowledge it is operating in looks much like a mesa. That is, in those sections of the knowledge domain that it knows about, an expert system gives a quite capable performance, frequently as good or even better than that of a human being. However, if it is placed in a situation even slightly outside of its area of expertise, the performance drops sharply—just like stepping off the edge of a mesa. Worst of all, an expert system has no knowledge of what it does and does not know; it gives no warning that its performance has dropped from expert levels to near-idiot levels. Unless the user realizes what has happened, the often nonsensical advice provided by the expert system in such situations can be considered as gospel.

Does this sound like the problem of country-bumpkin common sense applied to city-slicker environments (or vice versa)? Absolutely. The difficulty in both instances is that expertise in one environment may prove to be idiocy in another. And because an expert system is sublimely unaware of its own limitations, it may not be the complete answer in providing an android with robust common sense.

Is there any other way to implement common sense in an android? Researchers in AI have been considering this problem for years. One partial answer has been proposed by Roger Schank. He believes that the answer may lie in the development of storage modules he calls "scripts."

In Schank's view, all of us have a large collection of scripts stored in our heads that are designed to carry us through a variety of situations. Take, for example, the problem of eating in a restaurant. Our behavior, in his theory, is determined by a "restaurant script" that tells us the appropriate response to the likely events that occur in a restaurant. As long as real life follows the script, we don't have to think carefully about what we are doing—we can merely enjoy the food, company, and ambiance. Only when something unusual happens, something not covered by our inner script, do we have to consciously decide what to do and how to react; the rest of the time we can concentrate on something else.

What would such a script contain? Consider what happens when one eats at a restaurant. The diner walks in the door, sits down at a table, orders a meal, eats the food, pays the bill, and leaves. These are basic actions for the restaurant script. But this simple description is not nearly good enough, because eating at a restaurant requires interactions with other people that are not covered by this script. Let's try again.

A more detailed script might go something like this: the diner walks in, asks the host for a table, follows the host to a table, sits down, opens the menu, chooses a meal, and gives the order to a waiter, who later brings the food. The diner then eats the food, accepts the bill from the waiter, places an appropriate tip on the table, walks to the cashier, pays the bill, and leaves the restaurant. If we like, we can have the restaurant script call other scripts for some functions. For example, that portion of the script covered by "the diner eats the food" could just as easily call a secondary script (or scripts) that control exactly how the food is eaten. There might be scripts for "eating a sandwich," "drinking a beverage," and "eating a salad" as well as many other specific tasks. These subscripts might in turn call other scripts covering such details as how to handle cutlery and general table manners. By postulating a wide collection of interrelated scripts we can generate a "restaurant script" sufficiently detailed yet general enough to cover many situations.

Notice also that the revised restaurant script has more than one actor in it. There are now roles for each of the diner, the host, the waiter, and the cashier. These may or may not all be different people, but their identities at any given time (and thus the diner's reaction to them) are determined by the role they play in the script. For example, if the waiter is the person who shows the diner to the table, that merely means that the same actor moved from playing the "host" role to the "waiter" role as the script progressed. A role can refer to a person, an animal, or an object; even the restaurant itself is a role in the script above, as is the menu and the food. Furthermore, each role can have characteristics associated with it that help identify the various players in the script. These identifying characteristics must not be too restrictive, however, since otherwise the role identification would be rejected if the diner encountered a waitress instead of a waiter!

A script-based system keeps each script as a sort of template for actions and behaviors. When the current perception of the world matches a particular script template closely enough, a matching attempt is initiated to associate the open roles in the script with actors and objects in the environment. Once a particular player has been identified as the "waiter," for example, that person's behavior becomes predictable and the meanings behind his actions are considered to be understood, along with the proper reactions expected from the system (or, in this case, the diner). As long as the players in the script more or less follow the actions outlined, the script provides the correct actions and reactions and all goes well.

Scripts thus seem to provide a reasonably powerful technique for establishing general rules of behavior in many situations. There are some problems with this, however. One is the large number of scripts that are required. Just consider the restaurant problem, for example. Obviously, the script described is a generally useful one for many restaurants. But a diner wandering into a cafeteria by mistake would make several major gaffes by following this script. And if the restaurant chosen is a fast food restaurant, the diner would likely starve to death waiting for the mythical host to show the diner to a table!

Apparently, if Schank's notion is correct, people carry with them a large number of similar scripts that cover variations on the basic eat-in-a-restaurant idea. We might have an eat-at-McDonalds script, an eat-at-the-coffee-shop script, an eat-at-a-cafeteria script, an eat-at-the-deli script, and an eat-at-an-expensive-restaurant script. There might be a number of other variations on these as well to cover the variety of specific restaurants we actually patronize, and that make us able to cope with these various situations. It is possible, in fact, that much of the discomfort people often experience in a new environment stems from the lack of an internal script to govern their behaviors and actions in this setting. Thus, the first time a person eats at a fancy restaurant, he or she is likely to experience the uncomfortable "everyone is staring at me" syndrome; once the proper script for behavior in this new environment is absorbed, that discomfort fades away.

Scripts, and a similar notion called frames, offer an intriguing glimpse at the kinds of internal structures that might be part of a person's common sense, but these notions are still incomplete. We need to add the idea of inheritance to make this system come alive.

Inheritance comes from the concept of categorization. If I tell you that I have a pet cat named Delilah, you already know a great deal about her. Because Delilah is a cat, you know that she is a furry, warm-blooded mammal, with a tail and four legs, and that she probably likes to sleep in the sun and be scratched under the chin. How do you know this? Because these are common charactistics of housecats. Essentially, you have mentally created a specific instance, called "Delilah" of the general category of "housecat." As you learn more about her, the Delilah-instance gains more detail, but even initially you assume that this specific case of a housecat has the same characteristics of the more general category. You don't have to remember specifically that Delilah is furry and has a tail; you simply store that information—along with many other similar details—as part of the general housecat category. This system of allowing specific cases to assume the qualities of general categories unless specified otherwise is the basis for inheritance.

The power of this scheme comes when you learn more than a single instance for a given category. Suppose, for example, you also know that Bob down the street has a cat named Mittens. You can construct a similar cat-instance for Mittens, that again inherits the general characteristics of a cat. This inheritance scheme thus allows you to store the general details of a cat only one time, in the definition of the category, and not every time you learn about a specific cat. Only exceptions to the general characteristics, the characteristics that make Mittens and Delilah unique animals, have to be explicitly stored in each specific instance.

The notion of inheritance is a powerful one that has caught on in computer programming like wildfire. Originally a part of the specialized field of artificial intelligence, now modified versions have crept into a number of popular general programming languages. In such cases it has expanded into something called "object-oriented programming," and is implemented by languages like Smalltalk, and C++. Object-oriented programming simply means that a program's data consists of a number of objects that are related to each other—like the two instances "Delilah" and "Mittens" are related to each other by sharing a common category of "cat." Objects in such a programming language communicate by passing messages to each other. Each kind of object responds to a given message in its own fashion, causing it to do something useful and (possibly) generate new messages that get passed to other objects in the program. For example, if you have a cat-instance of Delilah and a dog-instance of Ferdinand, you might pass each of them a message: "Master is home." Their response to this message is determined by the kind of object they are, cat or dog. Thus the Delilah-object might lazily roll over on her back waiting to have her tummy rubbed, while the Ferdinand-object might come running up to Master, drop a favorite toy at his feet, wag its tail, and bark. While both receive the same message, their responses to that message differ greatly. Similarly, multiple objects in an object-oriented program may receive the same message from other objects in the system, but how each responds to that object can vary enormously.

In a general computer program of course, these objects are more likely to be something like a print-object, which receives messages of documents that are to be formatted and sent to the printer, or a calculation-object, which takes data as input and responds with a message giving the result of some calculation. Nevertheless, these less-exotic objects are considered to be just like a cat-instance within an object-oriented scheme of the world—though no doubt every housecat would be vastly offended by such a comparison.

There is a still-more exotic version of the notion of object-oriented programming in which the objects exhibit a certain amount of mutability. This kind of programming is called actor-oriented programming. The primary distinction between an "actor" and an "object" is that an actor may change its behavior over the course of a program, while objects do not have that flexibility. This increase in adaptability means that an actors-based program is usually similarly more adaptable. The only drawback to this is that actors may be somewhat more difficult to program than objects for this reason.

Object-oriented programming techniques differ greatly from those of standard procedural languages like Basic, Fortran, and C. The programming metaphor itself changes from one of a detailed list of instructions to one in which more-or-less independent instances of abstract objects interact among themselves to accomplish a goal. Programmers new to object-oriented programming frequently feel that they have lost some control of the program, as objects are created and destroyed based on messages passed among the objects. It should come as no surprise that few programmers can make this mental shift in perspective without some retraining.

To make the android have a modicum of common sense requires an ability to implement inheritance as well as scripts. With inheritance, the android can infer characteristics of objects based on what it knows of other, similar objects. A person naturally assumes that some new item that looks like a known object has similar characteristics to that known object. A plant that looks like a flower is expected to smell good—or at least have no smell at all; an animal that looks like a teddy bear is expected to be equally cuddly. Never mind that in real life the beautiful white flowers of the thorn apple smell absolutely putrid, or that the cute and cuddly koala can be vile-tempered; until we explicitly note these as exceptions, we assume that they are similar to our stereotypical understanding.

Inheritance is thus, in some sense, a matter of stereotyping. Stereotypes have gotten a bad reputation recently; the fact of the matter is, however, that stereotyping, when judiciously applied, is actually a very powerful means of coping with the enormous mass of our daily experiences. By lumping our experiences into categories we manage to avoid having to deal separately with each one—a saving of time and effort that prevents us from experiencing complete information overload. We use stereotypes when we lack sufficient information to draw unique judgments about the events, people, and objects we encounter. After all, if we see a pretty flower and assume that it smells nice, most times we are right; only occasionally is that assumption wrong. If our notions about what a flower smells like are wrong too often, we can always update our stereotypical idea of a flower to account for these errors; if our template for flower is only occasionally inaccurate, we are likely to just note the errors as exceptions to the general rule. By making these assumptions about new encounters, we save ourselves not only the effort of testing every flower to see if it smells nice, but also possible dangers. Suppose you hold a notion that pulling the tail of a cat is likely to be risky because you once pulled the tail of a housecat and were scratched as a result. Holding that stereotype in memory would certainly prevent you from testing this hypothesis on every cat you see, including the lion at the zoo, or the bobcat in the wilderness. Without a stereotype used to infer that the latter two behaviors were likely to be damaging to the self-esteem—not to mention fingers—life would be far more dangerous.

Inheritance thus offers one means of dealing with stereotypes for objects encountered daily. It is much the same as reasoning from specific instances to general principles. While not all the conclusions drawn are valid, ideally they are right more often than wrong. When we resort to stereotypes we are not thinking in the sense of logically working through a problem; we are merely assuming that experiences we have previously perceived are similar to those we perceive in the future. This, of course, is not true all the time, but such assumptions do help us cope with many parts of daily life.

In a way, scripts are also a means of stereotyping an experience. We assume that what occurs in a restaurant is similar to what has happened in other restaurants in the past. Most of the time, this assumption is valid. When it is not, we are generally quite surprised, and may not know how to react at first. While human beings may enjoy perceiving themselves as creatures of thought, much of the time we appear to operate on a fairly automatic basis, only really thinking about those few truly novel experiences that occur each day.

Providing the android with the ability to use and modify stereotypes is one more way we can enable it to cope with the world around it. Much common sense boils down to such pre-digested "truths." Avoiding bad luck by not allowing a black cat to cross your path used to be only common sense; now it is considered mere superstition with no factual basis. The line between the two can sometimes be so fine as to be nearly invisible.

Generalizing from specific instances is also one of the key characteristics of neural networks. Recall that they learn by example. By providing the network with an appropriate collection of training examples, we can control what kinds of lessons the network learns. Nearly all neural networks have this capability, and sometimes the generalizations they perform are not exactly what the designers intended.

Recently, a neural network was trained by Terrence Sejnowski and Paul Gorman to "listen" to sonar echoes from two different kinds of objects, an underwater metal cylinder, and a similarly shaped underwater rock. After training, the network was as accurate as the best human sonar operators, and could correctly identify incoming signals a bit more than 90 percent of the time. Sejnowski and Gorman then analyzed the network to see how it went about making its judgments about the sound patterns. In many networks, this analysis is done by looking at the values for the weights on the interconnections, determining which weights are stronger or weaker than average, and then comparing this to the input patterns to see which stimulus characteristics correspond to the very strong weights and which ones to very weak weights. Since many weights in any trained network are inhibitory (negative), the analysis must take this into account as well. For a strongly positive connection to a particular neurode, the analysis usually concludes that corresponding signal features arriving over that connection is a positive determiner—it should be present for the neurode to fire; in the case of a strong negative weight, the corresponding signal characteristic should be absent rather than present for the receiving neurode to fire. In the case of a sonar echo, this weight analysis technique becomes more complex. There are no obvious characteristics of the input data for the researcher to use in doing the analysis, so statistical and clustering techniques must be used to determine what the network is doing. Nevertheless, although the process can be tedious, the results of the analysis are often fascinating.

When Sejnowski and Gorman analyzed the sonar network, they were surprised at the results. The network had developed an unexpected strategy for choosing between "rock" and "cylinder." Every new input pattern was assumed to be a cylinder unless it had characteristics that changed the network's choice to rock. This result may not have been predicted, but (in a wonderful example of 20–20 hindsight) it makes perfect sense. Given that the problem is to categorize an input pattern into one of two choices, why not assume that all inputs are one particular choice, and thus only learn the characteristics that make an input not that choice? It's really both efficient and ingenious, and involves learning the minimum amount of information. It is not necessary for the network to know all the characteristics of "rockness" and "cylinderness"; it is only necessary that it know the difference between the two.

In a sense, this network developed its own "common sense" approach to identifying the sonar signal. Its solution to this problem turned out to be both efficient and elegant. Like many trained neural networks, this system was able to generalize from the specific training instances it was given, creating a powerful generalization from those instances. Neural network researchers are often surprised by the innovative techniques used by networks to solve problems.

While we may be surprised by the power of systems that can generalize and that possess a modicum of common sense, it is not the same as having the ability to do truly creative thought. Many people argue that no man-made machine can be truly creative, because creativity is not a process that can be expressed as a sequence of steps or a collection of rules. Even defining creativity itself is a non-trivial task worthy of a roomful of negotiators. My dictionary (the *American Heritage Dictionary, Second College Edition*) defines creativity as "characterized by originality and expressiveness; imaginative" and as "the act of creating" where creating means "[producing] through artistic or imaginative effort." Apparently creativity has something to do with imagination.

What does it take to be creative? There is no consensus on how creative minds arise, or even on what characteristics are necessary for a person to exhibit creativity. Certainly, however, there are at least two key elements to the act of creation. First, the person must be able to imagine the result: Even if the reality doesn't turn out exactly as planned, the creator must have previously imagined that he or she was going to make something. Second, the creative person seems to be able to perceive relationships between events and objects that are not obviously related. Essentially, this boils down to being able to link, in a novel way, concepts that have no obvious connection. The first characteristic enables the person to plan an act of creation; the second provides the means of solving the problems involved in implementing the plan.

The second of these characteristics implies that creativity has something to do with applying old concepts and techniques in new ways. Consider what this means for a moment. If you take a selection of paints and apply them to a blank canvas in a way that has never been done before, you are being creative. Of course, the artistic merits of the pictures you create in this way may vary greatly, but the point is that it is the act of doing something new, something that has never been done before, that marks your work as creative, not the esthetic beauty (or lack thereof) that results.

Is applying paint in a never-before-seen pattern sufficient to identify that act as creative? Almost certainly not. If that were the case, an android or a computer could simply perform an exhaustive search through all alternative applications of paint, and would thus become instantly creative. What makes the painting creative is more that a new means of expressing something has been developed, with that medium of expression being both similar to and yet different from previous expressions of similar messages.

Modern Western artists most often describe their creative acts with words such as "self-expression" and "creative tension." Yet, as Morris Berman has eloquently pointed out, such terms are an artifact of today's Western cultures, and are not relevant in a broader historical context. The art of the Middle Ages and early Renaissance in Western Europe, as well as that of most Eastern countries, displays an entirely different form of creativity in which self-expression and the search for answers falls beneath the recognition that the answers are clearly understood by the artist. The Western artist experiences conflict and dissension; the Eastern artist deals with passivity and understanding. Creativity, it seems, comes in many forms, and some of these are foreign to those of us with a Western heritage.

How then are we to determine when an act is creative and when it is not? The question is a serious one, particularly in light of a number of recent artistic "discoveries." Animals ranging from horses to gorillas have been trained to produce paintings; some of these paintings have been enthusiastically received by unknowing art critics as exemplifying a superior artistic sense. The product of the creative endeavor cannot be the sole determiner of the strength of the creative experience. Instead, so critics of these animal artists claim, there must be some *intentionality* behind the production of the work. The trainers of these animals frequently assert that their budding artists "enjoy" the process of creating a painting, and therefore their artwork falls into the realm of genuinely creative acts. While human artists who have adopted the "art is suffering" school may criticize such an assumption, there is no way to look into a horse's mind (or even that of a gorilla) to determine if the horse achieves some level of mental or spiritual satisfaction in its creative acts.

Creativity is a mystery in humans; how then should we judge it in androids? It is almost certainly true that we can expect to build androids that can paint pictures an art critic would call "artistic" or that can play music with an expressiveness that would fool most listeners into believing that a human was playing. But are such performances mere imitations, good enough to fool many perhaps, but underneath nothing more than mere shadows of truly creative acts? And how can we tell the difference?

Suppose for the moment we use an approach familiar in science, that of making an operational definition of creativity. This means that we define creativity by providing a collection of measurable or observable behaviors which, if achieved, we agree exemplify a creative act. In other words, for now we leave all indeterminate quantities such as "intentionality" and "self-expression" aside and concentrate on actions that we can verify.

One basic definition of creativity is that it occurs when an entity engages in deliberate acts of a novel character that produce a recognizable object or event. The actions must be intentional; they cannot be the result of happenstance to be creative (although the outcomes of the actions may not be totally predictable, the entity must have intended that *something* would result from its actions).

With a definition such as this we can see the basic qualities that android creativity must meet. First, the android must have generated a plan that includes the actions it performs; the actions must have the production of an object or event as their goal; and the result of the actions—either the product of the actions or the process of generating the product—must have some kind of novel character to it. How could we build an android that exemplifies all three of these traits?

The first characteristic, planning, is easy. As we have seen, androids must be able to plan merely in order to move about in their environment. Any android that can navigate successfully in the real world has ample planning capabilities for any creative endeavor. But we should not be too glib about even this ability, for successful planning implies the existence of a goal. Did you ever try to make a plan to accomplish something when you had no idea of what it was you wanted to accomplish? Even an android with a highly competent planning module must be able to envision a goal that it can then plan how to achieve. The key question here is whether an android has the ability to envision a novel goal. Given such a goal, an android would almost certainly be able to produce and carry out a plan to achieve the goal.

A similar difficulty exists with the third criterion for creativity, that of having some novel character to either the result of the plan or the process used in the plan. If an android envisions a goal of winding a mantel clock, and achieves that goal by executing exactly the same—or even very similar—sequence of actions it has used for the last twelve weeks, that action is not creative. On the other hand, if the android generates an entirely novel method of winding the clock—perhaps by using a mechanical egg-beater to turn the clock-key—that is certainly a novel, *creative* approach to the problem.

To provide an android with the ability to imagine is likely to be difficult, but to give it the ability to draw new associations may not be hard at all. An associative memory system—which we've already determined must be part of any successful android—works by associating new stimuli to stored information. While this kind of memory is not perfect and does not always make the correct association or recall the "right" concept, this very same lack of perfection occasionally means that new associations are drawn. For example, a person may look up at a cloud and say that it looks like his great-aunt Beatrice smoking a pipe; another may say that the same cloud looks like a basset hound with its tongue hanging out. While it's possible that great-aunt Beatrice really did look remarkably like a basset hound, it's more likely that the cloud actually has little in common with either one. Instead, some feature of the cloud formation reminds the observers of some other object, either Beatrice or a basset.

People draw such off-the-wall associations all the time. They are the basis of similes and metaphors. It is not necessary for a person to receive a stimulus that is accurate in every detail in order to make an association with something; this is one of the key characteristics of associative memories. While even a crude stimulus generally generates the intended response, it sometimes results in an odd reaction. The result of this is that an occasional random stimulus may spark the recall of a memory that has little to do with the original stimulus. A new association has been drawn—the cloud is like great-aunt Beatrice, or even great-aunt Beatrice is like a basset—and that is a major key to being creative.

Putting an associative memory into an android allows it the opportunity to make such new associations. Thus, the very feature that occasionally causes an associative memory to give the wrong answer is also the feature that gives it the ability to conceive of new associations. The skill that must be provided to the android is one of not discarding such mis-associations, and of recognizing when they can be applied to a current problem.

And yet, as Roger Schank has pointed out, the mis-associations cannot be just any kind of error. They must be near enough to the "correct" or traditional answer to make some sense and to allow a logical chain of reasoning to connect them to the existing situation. Schank's group at Yale has succeeded in developing a program that can explain simple stories with remarkable novelty. Their system uses a knowledge representation model of event explanations that is similar to the scripts described earlier. In this case, a given explanation is fitted with three kinds of situational contexts: (1) a list of situations in which the explanation directly applies; (2) a list of broader situations in which the explanation may not directly apply but may be relevant; and (3) a causal list of the events included in the explanation.

A simple example of this kind of knowledge structure might go something like this. The situation that is directly described by the explanation is that by mending a small tear in an item of clothing, a larger mending task is averted. A broader situation in which the explanation may also be relevant is that bad outcomes can be avoided if prompt preventative action is taken in advance; a second broad situation is that bad circumstances only get worse if not dealt with promptly. Finally, the causal list of events in the explanation is: A small tear in clothing is easy to fix. If a small tear is not fixed promptly, it grows larger. If it grows larger, it is much harder to fix.

Did you recognize this explanation pattern as corresponding to the proverb, "a stitch in time saves nine"?

Creativity in an android may be as much a by-product of its memory system as a deliberate design specification. Such properties are called emergent properties because their appearance is not directly part of the planned behavior of the system. When the other, planned characteristics of the system are implemented, the interactions between these subsystems may cause the android to exhibit new properties. I will have much to say about emergent properties in the next few chapters. For now, though, the combination of language, planning, and script capabilities needed to make a practical android may very likely result in creative behavior emerging in its repertoire of behaviors.

Now, how many androids do you think it will take to change a light bulb?

≧ 11 ≦

I Think, Therefore I Am—I Think?

What's on your mind, if you will allow the overstatement?

Fred Allen

In *2001: A Space Odyssey* the HAL 9000 computer system experiences a minor social malfunction that causes it to deliberately murder one crew member, to attempt to strand a second in a shuttle pod in space, and to turn off the life support systems of the several crew members who are traveling in suspended animation, thus killing them. Until this time, no HAL 9000 series computer has ever exhibited the slightest error; it is considered as close to perfect in operation as any man-made device can be. Is HAL the future of androids?

Alternatively, consider the robots of Clifford Simak's classic novel *City*. Here they have become far more than servants to humans; they also are friends, nurturers, caretakers—and mourners when mankind dies. They demonstrate warmth, creativity, and personality, as they truly become the inheritors of mankind's position on the world. They take it upon themselves to shepherd mankind's other loyal followers, the dogs, into the realm of civilized society, and rescue life on Earth from the blind malevolence of the insects. Is this the future of androids?

Or consider Isaac Asimov's positronic robots, which obey the Three Laws of Robotics: (1) No robot through its actions, or lack of action, shall cause any harm to any human being; (2) A robot must obey the commands of any human, except where that would contradict the First Law; and (3) A robot must attempt to prevent harm from coming to itself, except where that would contradict either of the first two laws. One of the inventors of the positronic brain says about them, "To you a robot is a robot. . . . But you haven't worked with them. . . . They're a cleaner, better breed than we are." And indeed, in his robot stories and novels, they are somehow better and nobler than mankind. Is this the future of androids?

Which of these scenarios is closest to the truth? Obviously, we must wait and see for ourselves what the future holds. Yet science fiction writers have presented us with a myriad of possibilities for their future as partners with us. Which will turn out to be the truth, if any do at all?

So far, we have viewed an android in a manner that is a bit like the three blind men studying an elephant: The one groping at the elephant's legs thinks it must look like a tree; the one clutching the elephant's trunk believes it is more like a giant snake; while the one being swatted by the elephant's tail thinks it most resembles a broom. Our study of androids has been like this. We have considered its various subsystems for vision, language, locomotion, and more; now it is time to consider how all these pieces might interact with each other to make a complete being.

It is a truism that a complex system is much more than the sum of its parts. Once all the pieces are pulled together, a synergy develops that makes them together capable of more than they are while apart. It is likely to be that way when we build an android as well. It will have eyes to see with, hands to manipulate its environment, and legs (or some other means) to move it about. Most of all, it will have a brain that is capable of learning from what it experiences—not as much or as easily as a person can learn, perhaps, but capable of learning nevertheless. Furthermore, it will have what no other animal on earth besides ourselves has: a rich, human language in which to express its thoughts. All these traits are essential characteristics of the android as we have defined it.

Is it necessarily true that androids will behave as people do? Of course not. One of the key lessons of neural network research is that while their overall behavior may be similar to that of a person, the methods they use to arrive at that behavior are often quite different from what a person would choose. For example, if a neural network is trained to distinguish visual patterns, most typically use their middle layer or layers as various levels of feature detectors. But the features these neurodes are sensitive to are not necessarily the "obvious" ones a person would use. Indeed, while they are usually quite similar to the human choice of a feature set, they also are likely to vary substantially from that set. It is very probable that a neural network brain in an android will have the same characteristics: similarity in behavior to a person, but not an exact duplication of a human brain.

The critical characteristic of an android is the fact that it must be made up of a substantial number of highly interconnected systems.

The locomotion system must be closely communicative with the visual system (for steering) and the higher brain functions (for path planning); the coordination system needs similar ties. The hearing and speech systems must similarly connect to the higher brain so that it can understand and generate language. In fact, the brain is the central connecting point for all the major systems in the android. It must assist with all their functions, and must monitor and react to problems within the entire android.

This monitoring function is an essential one, for a system as complex as this may have hardware or other failures that are not at all obvious to any nearby humans. This means that it must have at least some ability to detect and diagnose errors and problems for itself. Of course this is not at all surprising, since most complex electronic systems today have self-diagnostic modes to a greater or lesser degree. But in an autonomous android, operating more or less independently of humans, this implies that the system must not only have diagnostic capabilities, but also have the sophistication to decide what should be done about problems: Should it return to its base for repair, for example, or complete its current task, and then return? Only a system that understands both the goals of the current task and also the implications of each kind of failure mode can respond properly to such circumstances. For an android to operate independently, this is essential.

This implies that the android must have a sense of self-awareness, at least to a limited degree, and at least to the extent that it is aware of the conditions of its own body. Furthermore, it is likely to be desirable that the android have a sense of self-worth to keep it from taking foolish risks. After all, these devices will not be cheap to build, and no owner is going to want to see one of them deliberately walk off a cliff. The easiest way to set up an avoidance of injury is to provide the android with some strong inhibitions against performing certain acts that are likely to cause damage. In concert with this, though, must be some strong stimuli to get damage corrected as soon as possible—and the urge should be stronger the greater the damage to the android. What I have described here is the need to provide it with a sense of pain.

Think about the function of pain in a person. Pain doesn't exist for no reason. It is there to tell us that we have experienced something that damaged us, and its purpose is to restrict our actions long enough for our bodies to recover from that damage, and thus avoid inflicting even more serious damage on ourselves. Those few people who are born without pain are not lucky, as some might believe. Instead they must be on guard every moment of their lives, constantly checking to make sure they have not cut themselves, or broken a leg, or smashed a finger. It is entirely possible for such people to bleed to death and not realize until too late that they were badly hurt. Pain is actually a safeguard against this kind of damage, and we definitely want to include it in any android we build. It must be capable of sensing damage to itself, and it must strive to minimize those events and correct them as soon as possible when they do occur.

What about pleasure? Pleasure is less easy to justify than pain, because it does not have obvious diagnostic effects. However, in one respect, pleasure too may be a necessary subsystem. One of the interesting results of recent neural network research suggests that certain networks that are trained with positive reinforcement—praise, in other words—may learn somewhat better than those trained with negative reinforcement. While these suggestions are far from conclusive at this time, they are certainly supported by any number of psychological studies with people and animals. Assuming that this is true, it seems logical that for some kinds of training it will be very useful to be able to train the android's neural networks as forcefully as possible. Building it so that being praised is a desirable state, while being not-praised is an undesirable state, is a primitive equivalent to a sense of pleasure. In essence, the android would feel pleasure in a job well done, and disappointment when it failed at a task.

All this may sound highly anthropomorphic and suspect, particularly to those people who have not yet conceded even that animals can have feelings. Yet anyone with a close relationship with their pet dog or cat is well aware of their moods and feelings. While it may not yet be accepted truth in all circles, there are few loving pet owners who doubt that their pet experiences them. And no one who has ever been in a veterinarian's office can doubt that animals experience pain. The real anthropomorphic fault is the arrogance that permits humanity to believe that their feelings are superior to those of other animals. Chimpanzees and humans have nearly identical genetic codes, yet we label them as "only" animals, and consider ourselves as greatly superior. We can get away with that with chimpanzees, for they cannot express themselves with our language. I suspect we will not be able to get away with this with androids, for they will be created knowing how to say what they feel.

So what we have now is an android that can experience, at a minimum, a feeling of pleasure at a job well done, and a feeling of pain or sadness or disappointment when failure or physical damage occurs. Notice that this does not make the android a particularly emotional creature. Rather it provides it with what might be called a basic survival kit of emotions, just enough with which to get by. Even compared to *Star Trek*'s Mr. Spock, these androids are likely to be pretty stiff creatures indeed.

If androids are likely to have at least this minimal version of human feelings, are they likely to have other human problems as well? Can they ever become mentally ill? Or is it possible that, like HAL in *2001*, they may prove to be just the slightest bit socially maladjusted?

Consider the explanation offered (in *2010*, the sequel to *2001: A Space Odyssey*) for why HAL malfunctioned. In essence, he—one cannot help calling this computer "he"—was provided with contradictory instructions, part of which forced him to lie to the human crewmembers. His design and training gave him no experience with lies, so this set up an irreconcilable conflict within his logic circuits. The result? To avoid having to lie, he eliminated those to whom he would have had to lie—the human crew members. While being pure science fiction, this explanation is science fiction at its best because it is entirely plausible. Let's see why.

Presumably, any advanced android—or computer, as in HAL's case—will have higher logic centers. These logic centers will most probably implement a combination of neural networks, expert systems, and other AI reasoning techniques, if only because these have so far proven to be the most effective tools we have yet found that can perform problem solving. All these tools have some potential pitfalls, however, chief of which is that they possess only a limited area of expertise. In the case of a neural network, that expertise is limited to whatever domain of knowledge on which it has been trained. While it may be able to extrapolate, generalize, and otherwise extend its capabilities through some judicious guessing, as we all know, such extrapolations do not always make sense in the real world. It would be a bit like a successful American business executive being shipped to Japan and being ordered to quickly become equally successful there. While the native intelligence and experience of the American help some, methods of doing business in Japan are sufficiently different from practices in the United States that the only result the American is likely to achieve is failure—unless, of course, the executive is given some additional training in how the Japanese do business.

Expert systems have similar drawbacks, and perhaps even more severe ones than neural networks. Expert systems have long been known to exhibit the "mesa effect" in performance runs. As long as they are asked to solve a problem within their area of expertise, they do quite well. But as soon as the problem is even slightly out of that area, their performance drops abysmally—just like stepping off the side of a mesa. Worse, expert systems generally don't know that they don't know. (With a neural network, in contrast, there is often some kind of confidence level that tells us how sure the network is of its answer.) That is, given any problem, they blindly try to solve it, whether or not it has anything to do with their area of expertise. Thus, the response from an expert system can be either truly expert or truly nonsensical, and it is up to the user to ensure that he or she knows which is which.

Given these limitations in both neural networks and expert systems, it is easy to see why HAL might be vulnerable to a few well-placed lies. If he had no experience during training with lies (and why would anyone train a computer to lie?), the expert systems that controlled his judgment might not realize that they were operating in an area that was outside his domain of knowledge, and thus HAL might easily walk (figuratively speaking, of course) right off the edge of that mesa.

Before androids can be turned loose on the general public, they must have some training in dealing with humans. And this means they must know that humans do not always tell the truth, that some are less than honorable, and that, just perhaps, it might not be a good idea to obey the human's instructions to blow open the bank vault. They must have instilled within them some level of moral code, in other words. Asimov's Laws of Robotics make an excellent starting point, but they are not sufficient. There is nothing there, for example, that would stop the bank robber from ordering the bank's android guard to open the vault and giving him the money. And, from the second law, the android would have to obey as long as the action caused no harm to any person. Clearly additional constraints on the android's actions are needed, such as obeying all the laws of the land (but what happens when the laws contradict each other?), or protecting the property of humans as well as their physical beings.

There is one more important reason why we need to consider the actions of the android as a whole system, rather than just the sum of its pieces. By the time such a device is constructed, it will necessarily be one of the most complex machines ever invented. Such complex systems are often inherently unpredictable because of the vast number of ways the different subsystems can interact with and affect each other. It may well prove to be the case that an android is actually a chaotic system.

Chaos is a term used to describe systems that appear to be predictable in the short term, but actually are unpredictable over any extended period of time. The best example of a chaotic system, one that touches all our lives, is the weather. Has it surprised anyone that, after decades of efforts with the most powerful computers in the world, meteorologists still have great difficulty predicting the weather more than a day or two ahead? And long-range predictions of a week or a month or a season are more likely to be probabilistic estimates than accurate forecasts. Many scientists today think that this is as good as weather prediction is ever going to get, all because the weather is a chaotic system.

Weather is chaotic because it is very complex and because tiny measurement errors of the state of the weather *now* accumulate quickly and result in huge differences in the predicted states of the weather at times in the future. And the further off those future times are, the greater the divergence is. For example, a weather monitoring station may measure the local temperature at 59°F (15°C), the wind to be 5 miles per hour (8 kilometers per hour) from the northwest, and the sky to be sunny. But those measurements, no matter how carefully done, are likely to have errors associated with the accuracy of the instruments. The actual temperature might really be 59.1305°F, and the wind might really be blowing at 4.832 miles per hour. Every weather monitoring device, every measuring instrument in the world has some kind of limit to its accuracy. Furthermore, we do not and cannot have measuring instruments at every possible location in the world. Thus when the weatherman receives a report that it is 59° in San Diego, California, what the report really means is that it is approximately 59° at the location of the thermometer in the San Diego weather monitoring station.

When all these approximate measurements are plugged into the computer simulations of the atmosphere, they generate a pretty good forecast for up to 48 or 72 hours ahead. After that, the accumulated errors that result from treating approximate mesurements as exact begin to cause unreliable results. If the San Diego station reported a temperature of, say 59.5° instead of 59°, little difference would be noted in the weather forecasts for a time. But sooner or later that single tiny difference would extend itself until the new forecast had little relationship to the previous one. Changing only a single measurement such a small amount means that the discrepancies would probably occur months or years ahead. But when you consider that there are thousands and thousands of monitoring stations, each offering their *approximate* measurements of the current state of the weather, you begin to see how the errors accumulate so quickly.

Chaotic systems do not necessarily have to be hugely complex; they can consist of as little as three planetary bodies interacting with Newtonian gravitation and mechanics. However, all chaotic systems have the characteristic that small changes in their current conditions eventually result in hugely divergent resulting behaviors. No one can predict precisely how a stream of water tumbles down a rocky streambed. While physicists thoroughly understand the principles that control the water's overall behavior, they cannot hope to predict which drops of water end up splashing on the rocks at the side of the stream, and which merrily make their way downstream to join the ocean.

While overall global behavior is predictable in the short term, in the long run, the stream—like the weather—behaves in a manner that is entirely its own.

There is currently some discussion that certain neural networks might be wholly or partially chaotic. It is certainly true that taking a new set of random starting weights on the network's connections can have a huge impact on how quickly the network learns, or occasionally even whether it can learn at all. When we consider the truly complex neural networks that are bound to be part of a working android, it is easy to imagine how their complicated interconnections might result in some chaotic aspects. Perhaps this accounts for some of the inherent unpredictability of human behavior as well.

The point here, of course, is that with so many complex systems interacting with and depending on each other, we may not be able to completely predict what androids will do all the time. That does not mean that they will become instant ax-murderers or metallic felons because it should be reasonably simple to include strong, unbreakable inhibitions against such behavior. But it does mean that their actions will not be totally predictable.

You may find that your android butler has some behavioral quirks that you have to learn to live with. Perhaps it always harumphs before requesting further instructions, or perhaps it wipes its feet whenever it comes indoors, even when there is no doormat there. Or perhaps its speech patterns are formal and precise, except when it slips into a strong Southern drawl on some words or phrases. Whatever these quirks may be, you will probably not be able to train them completely out of its system because they will result from the interaction of all its complex systems, not the specific outputs of one system or the other. In fact, you may find such eccentricities somewhat endearing, just as I do when my cat insists on sitting in the middle of the morning paper I am trying to read. Oh, you may grumble and complain a bit, but you get used to such oddities in other people and pets, don't you?

Furthermore, these behavioral patterns will almost certainly vary from individual to individual. Because the quirks are caused by tiny variations in the structure and interaction of the subsystems composing the android, they are unlikely to be exactly duplicated in any copy. Thus, if we replace last year's android butler with the latest model, we might well exchange an occasional harumph for a Texas twang now and again. In a true sense, the androids we build may literally have more of a personality than we bargained for.

It is time now to come back to the question that was deferred at the beginning of this book. Is it possible to construct an android that is intelligent in a real sense? Does the android have a mind?

Eloquent and passionate arguments volley back and forth on this issue, with both sides of the argument exhibiting little patience for the opposing view. As charges and countercharges flow back and forth between the debators, neither side has demonstrated much tolerance for the other's perspective. The issue is a hotly emotional one, and few people refrain from taking sides in the debate. I also have a position in this regard, which is summarized below.*

Consider first the possibility that intelligent machines cannot be constructed, and that at best we can build shallow imitations of human beings. What arguments support this notion? The most compelling argument on this side of the debate was put forth by John Searle in 1980. He presented what has become known as the Chinese Room problem.

Searle's argument goes something like this: Suppose I construct a room and put a person in the room who knows nothing of the Chinese language. In the room are a collection of symbols—Chinese ideographs—as well as a set of rules for how to associate an ideograph with an English sentence. I then can pass the person a Chinese text and, by following those written rules, the person in the room can convert the Chinese text to English. If the rules are complete and correct, the text is translated from Chinese to English. In spite of this apparent "understanding" of the translated text, the person in the room needs to know nothing at all about the Chinese language: He is merely blindly following the rules that I provide, swapping meaningless symbol for meaningless symbol. (In fact, to make the situation even more compelling, a person who knows neither Chinese nor English could be placed in the room; in this case, the symbols at both ends of the translation are meaningless gibberish.) The person in the room has no true understanding of the task being performed. Searle concludes with the argument that "intelligent" machines are just like the person in the room; they blindly swap symbols, following some externally provided set of rules, but have no true comprehension of what they are doing.

As might be imagined, the Chinese Room problem provoked immediate and strong controversy. A large number of counterarguments have been presented, ably summarized by Raymond Kurzweil in *The Age of Intelligent Machines*. The most persuasive of these, I believe, is

*If you are interested in reading about these arguments in more detail, I recommend the following books: For the "mind is not computable" argument, see Roger Penrose's thoughtful discussion, *The Emperor's New Mind;* for the "intelligent androids are inevitable" argument, see Hans Moravec's *Mind Children* or Raymond Kurzweil's *The Age of Intelligent Machines*. Complete references are listed in the Suggested Reading list at the back of this book.

that Searle is incorrect in his assessment of the performance of the Chinese Room, and that in fact the system does now understand Chinese. The key, however, is that the "system" does not consist of the man in the room, nor of the papers with the rules and symbols written on them. The system is really the man, plus the papers, plus whatever else is used to make it able to do the translation. The man acts merely as the "central processing unit" of a rather uniquely defined translation system—and no one believes that the CPU of any computer is intelligent. What makes a person—or a computer, or a neural network—intelligent is its basic processing ability plus the knowledge that is processed or used to process information. When the processor is combined with the rules and knowledge, then the system *as a whole* has a comprehension much greater than the sum of the parts.

Another issue must be considered in this argument. As we know from experience with machine translation efforts, human languages cannot be translated by merely transliterating symbols from language to language. Rules can be developed that effectively convert the syntax of a statement, but these do not suffice to convey the meaning in the second language. Syntax does not provide complete meaning behind the original text. Instead, syntax must be supplemented with the semantics of the text, and those semantics must themselves be converted to the differing semantics of the second language. Just one simple example illustrates the importance of this point. In English, something that is slightly obscene is frequently described as "blue"; in Spanish a similar characteristic is described as "verde." A direct translation of "verde" to English is "green." But, depending on the semantics and context of the usage, the *intent* behind the use of the word may more accurately be translated "blue" rather than "green"; if "green" is used in an English translation, much of the message behind the text is lost. How is the translator to know which to use? It depends on the semantics, the message behind the sentence to be translated. At the least, this implies that any successful Chinese-Room translator must have sufficient world-knowledge, cultural-knowledge, and semantic-knowledge of both languages to be able to determine whether to translate the word as "blue" or "green." And if the system as a whole has this level of translation ability, it becomes extremely difficult to argue that it lacks an understanding of the language.

The system-as-a-whole argument is also very persuasive to those of us who work in the field of neural networks. If you consider an individual neurode in a network, you realize that this little "mini-processor" can most kindly be described as "dumb as dirt." It does absolutely nothing that any sane person would call smart. It merely responds to stimuli in a predictable, although nonlinear, fashion and transmits its response to lots of other, equally dumb neurodes. Nowhere in a neural network is there a single element that has any overt information about the kind of problem the network is trained to solve, nor is there anything that constrains a particular neurode to only working on, say, translation problems, rather than vision or speech or locomotion or any of a host of other "intelligent" behaviors. In spite of this, however, neural networks have already proved themselves highly capable at solving a large number of classic AI problems, sometimes displaying an astonishing level of competence. But it is not the individual neurodes that exhibit this expertise, any more than the individual neurons in the brain have any profound intelligence. It is the total cumulation of the actions of these individual, dumb devices that displays intelligent behavior. Only when the individual neurodes—or neurons—unite and interact does the system *as a whole* become more than the sum of its parts.

Intelligence in my view is not a place or a physical structure; it is a process that emerges from the cooperative behavior of some number of complex subsystems. An android that walks is not intelligent; one that talks is not necessarily intelligent; nor is one that plans or translates or autonomously learns. But an android that can perform all these actions *is* intelligent in a way that cannot be measured by considering only the individual parts of the android. Intelligence is *emergent* behavior: It is displayed only in the complex interaction of individually not-very-smart pieces and subsystems.

You cannot look at the brain of a human being and expect to find a section labeled the "seat of intelligence." The brain controls walking; does that make it intelligent? No, because any number of animals, such as insects, can move about in sophisticated motions, but no one would call insects intelligent beings. The brain controls vision; does vision make a person intelligent? Surely not, for then those people who are blind at birth could not be intelligent. Similar arguments exist for all the large-scale behaviors that characterize human beings. But if something that walks, talks, sees, hears, and exhibits other such individual behaviors isn't intelligent, what is?

Certainly language is an intelligent behavior. One of the reasons AI practitioners have placed such enormous emphasis on natural language processing is because it seems to be a key to higher intelligence. But this is a capability that is already proving achievable, although not yet in the sense of full understanding of all the complexities and implications of human language. Nevertheless, it seems highly unlikely that we will not be able to develop androids that can converse with us "intelligently" about any number of subjects. If we use language as the determining criterion for intelligent behavior, then it is certainly true that androids will be intelligent.

Language is a very anthropomorphic criterion for intelligence, however. Because we do not understand the languages of whale-song or dolphin-whistles or baboon-cries, it is very easy to consider these sounds to be mere animal noises. Are they that or something much more? We simply don't know at this stage, although it is difficult not to recognize that communication at some level takes place when baboons shriek warnings or whales sing.

In *The Emperor's New Mind* Roger Penrose offers another argument against intelligent computers. He argues forcefully that computers can never be intelligent because the mind performs actions that are not computable. To understand this argument, we need to digress a bit and talk about the underlying concept of all digital computers, the Turing machine.

A Turing machine is an abstract model of a computer that is made up of a recording tape and a processor. The processor has a list of instructions to execute such as "move forward one number," "halt," "read the current tape value," "move back one number," and "write a '1' on the tape at the current position." It can be shown that anything that a digital computer can compute can be accomplished by a Turing machine; thus, a Turing machine can execute any computable algorithm. The vast majority of real-world problems can be solved by a Turing machine, but there are others that are not computable in this fashion. One characteristic that makes an algorithm not computable is if a Turing machine can never reach the end of the computing process—if it requires an infinite number of steps to execute to completion. For example, the computation of the value of π to the last decimal place is not computable, because π is an irrational number that never terminates. Another problem that is not, in general, computable is determining whether a new algorithm is itself computable.

Computability of algorithms is a well-understood concept in computer science, and it is true that digital computers can only process computable algorithms. Any problem that is not computable—such as determining in general whether a problem or class of problems is computable*—cannot be performed by a computer. Unfortunately, however, human beings can easily solve a number of problems that are well known to be not computable. To Penrose, this means that

*While it is impossible *in general* to determine whether a problem is computable, it is entirely possible to make such decisions in specific cases.

digital computers can never have a true mind, and never be intelligent.

Penrose's argument is strong and persuasive when expressed in its full scope rather than in the brief form I present here, but it rests on two flawed assumptions. The first is that it is essential to build an intelligent system using only digital computers, and the second is the same problem Searle's argument encounters, that a system's capability is the sum of the capabilities of its parts. I have already discussed this second problem. In the first regard, the computability argument may be true; it may indeed be impossible to construct an intelligent computer. (In fact, my personal belief is that Penrose is completely correct in taking this position.) The fact is, however, that issues of computability do not hold up for systems that are composed—in whole or in part— of neural networks. Neural networks are not Turing machines; therefore the computability of an algorithm with a Turing machine is irrelevant when discussing the capabilities of a neural network.

What are the limits of performance of a neural network? Frankly, no one knows. The theoretical underpinnings of these systems are not yet understood, although many talented and capable scientists are working hard at developing the insights to understand these devices more fully. But even with the present lack of such a theory, I can very confidently assert that no theory will ever prove that neural networks are incapable of being intelligent. Why? Because we have at least one well-accepted example of an intelligent and capable neural network mind: that of a human being. Any theory that claims that neural networks can't be intelligent must also claim that human beings can't be intelligent—in which case the entire issue becomes moot. All we are after really is an android that is as smart as a person; if we change the definition of "intelligent" to exclude people, that's all right: We simultaneously change the definition of the goal of the android.

An intelligent android mind necessarily includes some neural networks as part of its overall system. It also must include a considerable amount of symbolic computing, expert systems, signal processing, and ordinary digital computers. Where will its mind reside? Like the human brain, the android brain will not have a specific place to which we can point and say, "This is where the mind is." The mind of an android, like the mind of a human, will emerge only when enough of the individual subsystems begin communicating among themselves to institute this mysterious quality.

In its own way, the birth of the android mind will be as mystical and awe-inspiring as the birth of a baby—and just as perplexing to philosophers and scientists alike.

≧ 12 ≦

But Is It Alive?

The chicken probably came before the egg because it is hard to imagine God wanting to sit on an egg.

Unknown

Mary Shelley wrote about a mad scientist who thought he could create life from lifelessness. The classic *Frankenstein* is arguably the first fictional expression of the layperson's fear of science and technology. Dr. Frankenstein's attempts to create a living human being proved fruitless, however, for he was never able to invest his creature with a soul. What about the android we have been building so carefully? Will it be a mere machine, or will it cross the very fuzzy boundary between living and nonliving—and how can we tell?

Many may have an instinctive reaction to this question that says, of course such a creation cannot possibly be alive. It is a mere machine, not a living being, after all. Why should we suppose that it will have the properties of life? The real question here, however, is not "Will it have the properties of life," but rather "What *are* the properties of life"? We need to spend a brief time exploring this question.

Suppose we try to define an animal. How do I know that my pet cat is alive, as opposed to not-alive? Well, she is a complex mechanism that operates on a more-or-less continuous basis. She breathes in oxygen, eats food (especially if she can steal it from my morning glass of milk), sleeps (constantly), moves (when it pleases her), and generally reacts to the world around her. I cannot say she can reproduce herself since she was spayed as a small kitten, but she would have been able to do so without that human intervention. Her body is a complex machine that is based on carbohydrates, and the individual cells within it are controlled by the DNA of her genes. She is a living animal just like the billions and billions of other living animals on this planet.

But are these characteristics necessary to the definition of life, or are they merely functional by-products of life? What distinguishes an animal from any other object in the world? We cannot insist that

Figure 12.1 The mysterious monolith of 2001 *and* 2010 *exemplifies an unanswerable question: Is it a mere machine, or is it somehow alive? (Photo courtesy of Turner Entertainment Company. Copyright © 1968 Turner Entertainment Company. All rights reserved.)*

respiration is a requirement, for we have discovered living animals in the anerobic environment surrounding deep sea vents. Certainly all living beings ingest food of one sort or another, but then so do machines. The purpose of ingesting food is to permit a chain of chemical reactions to occur within the animal (or plant), the result of which is that energy is released for use by the system. With this in mind, doesn't a vacuum cleaner similarly "ingest food" when it feeds from the electrical outlet in the wall? Of course plants and animals eliminate their waste products, but so does the vacuum cleaner when it heats the air around the motor and generates a roar of noise. The vacuum cleaner merely is more efficient at its digestion than, say, a cat because it directly ingests and eliminates energy.

Nor can we insist on sleep as a requirement for life, since it appears that not all animals sleep. All known *higher* animals appear to need to sleep, but it is not clear that all insects do. And does a jellyfish sleep? Besides, is not the period the vacuum cleaner is shut off a period of sleep? Certainly the vacuum cleaner has a limited "life span" that is shortened if it is operated continuously.

Mobility also cannot be what distinguishes an animal, for many animals are as rooted in place as a tree—corals and sponges are two examples. But animals do react to their environment in specific ways. They respond to changes in heat and light and react to the presence or absence of other life forms. For example, a cat bristles and hisses when a dog is nearby; the dog similarly barks and gives chase in reaction to the cat. Even a coral polyp responds to various stimuli by withdrawing into its stony home. Unfortunately, nonliving objects also exhibit some sophisticated behavior patterns in response to changes in the environment. A piece of litmus paper changes color in the presence of a base or acid, for example. While such behavior is exceedingly simple, so is the behavior of a sponge.

It is nearly impossible to isolate the distinction between living and nonliving based solely on behavioral phenomena. This was well illustrated in the mid-1980s when Valentino Braitenberg published a small gem of a book, *Vehicles.* In this book, Braitenberg demonstrated that many complex social behaviors can be simulated by cars that consist of only a few simple sensory receptors and a motor. The simplest of these vehicles have two sensory receptors—a device sensitive to light, for example—at the front, positioned much like headlights on an automobile. One version has each sensory receptor connected to the drive motor for wheels on the opposite side (left or right) as the receptor; the other version has each receptor connected to wheels on the same side as the receptor. Figure 12.2 illustrates the wiring of these two kinds of cars. Suppose these are indeed light-sensitive receptors, and when they are stimulated by light, they cause their connecting motors to speed up. Now place one of each kind of car on a flat surface with a light bulb somewhere ahead of them. What happens?

Consider the car in which the receptors stimulate the motors on the same side of the car (car (a) in the figure). If the right-most "headlight" receives a stronger stimulus (because the light ahead is somewhat to the right of the car's current position), then the right-hand wheels are made to turn faster than the left-hand wheels. The result? The car turns *left,* away from the light. If the light is on the left side of the car, the left-hand wheels are stimulated more than the right-hand wheels, and the car turns right, also away from the light. An observer unaware of how the cars are constructed might note that this car seemed to be *photophobic*—afraid of the light, in other words—because it constantly tries to flee whenever the light appears.

What about the other car, the one with its connections crossed? In that case, when the light is on the car's right side, the *left-hand* wheels are made to turn faster, so the car turns to the right, toward the light. The opposite happens if the light is on the car's left side—the right-

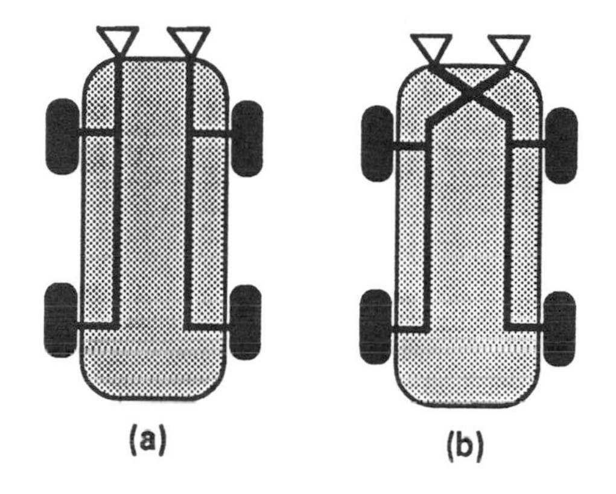

Figure 12.2 Two of Braitenberg's simpler vehicles. In (a) the photoreceptors stimulate the wheels on the same side of the car; in (b) the receptors stimulate the wheels on the opposite side.

hand wheels are stimulated more, and the car turns left. In effect, this car might be considered *phototropic*—attracted to the light—by the same observer.

There is another possible interpretation of these behaviors. In the first case, the car always turns away from the light, and tries to stay as far from it as possible. In effect, this car is a coward, always responding to the light by running away. The other car always turns toward the light, and must eventually run into it head on. It is possible to consider this car as an extremely aggressive beast, that always tries to ram its enemy. Thus, a simple wiring change converts the coward into the aggressor.

Braitenberg's vehicles exhibit much more sophisticated behaviors than shown in this simple example, however. The book presents more and more detailed versions of the cars that exhibit characteristics that, if only they were observed in an animal, would be considered part of rather complex social behaviors. The most sophisticated versions appear to behave with a sense of hope and optimism. They can predict the future—though, like people, their predictions are not always accurate—and they can learn from their mistakes and the accumulated wisdom of their predecessors. In many ways indeed they seem too animal-like—too *human*—to be mere machines. And yet their "mental makeup," the wiring and connections that constitute their controlling system, is still extraordinarily simple compared to the complexities of their behaviors. Because these behaviors are observed in devices we ourselves construct out of wheels, and wires, and bits of metal or plastic, we contend that they cannot be alive. But is that really true?

It is possible that the one characteristic that distinguishes living from nonliving is the ability to reproduce oneself. Animals and plants, if they have nothing else, certainly can do that, and they generally do so with alacrity. The urge to reproduce and perpetuate one's own kind seems to be an omnipresent biological imperative that crosses all species, whether plant or animal. In fact, an extraordinary amount of a plant or animal's energy is usually devoted to the production of the next generation of beings. There is even a standard joke in biological circles that says that a baby is a gene's method of reproducing itself. Given the elaborate reproductive strategies employed by any number of species, this does not sound all that farfetched.

Biological systems reproduce by first generating copies of their genetic code, the complex molecule that contains the fundamental instructions needed to construct a copy of the system. Sexual reproduction—the most common form among today's plant and animal species—requires the participation of two members of the species, which commingle their genetic codes to produce a third, unique version for their offspring. Asexual reproduction only requires a single parent, which then copies its own genetic code into its offspring. This is the original method of reproduction, and it is still used by some plants and animals, mostly of the microscopic variety. Sexual reproduction is the big success story of life on earth, however, because sexual reproduction permits evolution to take its course much more efficiently.

Animals reproduce their genetic structure by splitting the helical coils of their DNA into two matched half-coils. Each rung of the half-coils is then completed, one step at a time, until there are two versions of the original coil. Ideally, this copying occurs with no errors; in fact, however, several errors can occur during this process. By permitting the two distinctly different genetic codes of the parents to mix in the offspring, still more genetic scrambling occurs, resulting in a more dynamic and responsive species, one that can take advantage of environmental changes by evolving into new forms.

The rules that control how two genetic codes combine, and even how they accidentally make mistakes in combining, are quite simple. Many researchers have begun applying similar rules to practical problems in order to make systems capable of learning and responding to environmental changes that occur over time periods longer than the lifetime of individual systems. This field of study is called genetic algorithms, and it has begun to prove fruitful in some real-world problems.

In genetic algorithms, adaptation of a system to an environment is partially the result of probability, partially the application of specific algorithmic rules, and partially the result of plain dumb luck. Over the course of many generations, however, the results can be striking indeed. Let's take a few moments to see how genetic algorithms work.

Before a problem can be processed using genetic algorithms, the first step is to come up with a "genetic code" that can be used to signify the answer. For example, suppose the problem is to produce the word "good." A reasonable genetic code for this problem consists of strings of letters, each four characters long. Initially, we begin with some number of example strings that we can generate randomly. These correspond to the genetic codes of individual population members; the task is to produce an individual with a genetic code equal to "good." To make the example relatively simple, we restrict the letters permitted in any string to those in the desired word: "d," "g," and "o"; in general, however, no such restriction is required. We also assume that we have a fixed population size of six individuals in each generation. Again, in a more general case the size of the population may vary.

To begin, the genetic codes of each individual in the initial population pool, generation zero, are randomly set. An example of such an initial population is listed in Table 12.1.

Table 12.1. The codes of generation 0.

Individual	*Code*	*Fitness*	*After Copying*
1	gdod	0.75	gdod
2	gdog	0.50	gdog
3	odoo	0.25	odoo
4	dgdg	0.00	godo
5	godo	0.50	godo
6	dogg	0.25	godo
Total Fitness		**2.25**	

The fitness column of Table 12.1 shows the relative closeness of each individual code to the desired code of "good." For this simple case, the fitness simply allocates a value of 0.25 for each position that has the correct letter. A genetic algorithm begins by reproducing each code according to its relative fitness to survive (as determined by the fitness function). In this case, a fitness of 0.75 indicates that the code concerned has a 75 percent likelihood of surviving to reproduce; a fitness of 0.00 indicates that this code has no chance to reproduce before its "death." But since even the most well-adapted individual is also subject to accidents, random chance determines which ones are actually able to reproduce. Statistically, we expect that code number 1 will reproduce three times as often as number 3, but each individual case is strictly determined by chance. In this case when chance plays its necessary part, only four of the six codes actually survive to reproduce: numbers 1, 2, and 3 each make one copy apiece, and number

5—a very successful reproducer—makes three copies. Even though code number 1 is the most fit of this generation, it is not the one that reproduces the most. While the dice are loaded in number 1's favor, chance determines that in this particular case number 5 is more successful.

This is only the first step in producing the next generation of genetic codes. The surviving copies now enter the mating pool where they randomly select mates (or have mates selected for them). Once mates are selected, the genetic codes of the mating pairs are scrambled following a procedure called crossover. This means that at a specific location along the genetic string of each individual in a mating pair, the genetic codes are exchanged. For example, suppose the strings are "a-b-c-d-e" and "*f-g-h-i-j*," and the randomly selected crossover site is immediately after the second letter in each string. The resulting strings (after crossover) are then "a-b-*h-i-j*" and "*f-g*-c-d-e." In a four-character string, there are three possible places where the strings can cross, after each of the first, second, and third characters. The crossover site is randomly chosen for each mating pair, and the resulting code established. Table 12.2 shows these results. In the table, the entries with the same crossover site are the mating pairs.

Table 12.2. Generation 1.

Individual	*Code*	*Mate*	*Crossover Site*	*Final Code*	*Fitness*
1	gdod	godo	2	gddo	0.25
2	gdog	godo	3	gdoo	0.50
3	odoo	godo	1	oodo	0.25
4	godo	odoo	1	gdoo	0.50
5	godo	gdog	3	godg	0.50
6	godo	gdod	2	good	1.00
				Total Fitness	**3.00**

After only a single generation the desired genetic code of "good" has been achieved. You should notice as well that the overall fitness of generation one is significantly greater than the overall fitness of generation zero shown in Table 12.1. (By the way, this example is an honest one, with all the genetic codes, survival populations, and crossover sites determined by rolling dice to make the necessary decisions.)

Genetic algorithms can be more complex than this simple example, of course, but it does illustrate the power of the technique. If we randomly create four-character strings using the letters "d," "g," and "o" we find that only one out of a possible 3^4, or 81, combinations is correct. Taking six combinations at a time, and allowing repeats within each collection of six, we expect a probability of including "good" of only about 7.5 percent on each combination of six strings. It takes seven to nine such attempts before there is even a fifty-fifty chance of including "good" in one of them. In spite of that, the genetic algorithm technique finds the correct combination in the example above after only a single generation!

This leads to an understanding of the fallacy of saying that people are too complex to have occurred by the happenstance of evolution. That is indeed true—but then evolution does not occur by happenstance alone. It is true that chance plays an important role in determining which individuals will survive to pass along their genes, and also in deciding where genetic copy errors will occur during the reproduction process. But the power of genetics—and genetic algorithms—is such that changes are forced along appropriate paths much more quickly than random chance alone justifies. And just look at the difference that adding the ability to crossover during reproduction makes in the speed of changes. With asexual reproduction, significant genetic changes occur only when there are errors in copying the genetic code itself—a very rare event in real life. But when a mate is added along with the possibility of crossover, genetic scrambling is far more efficiently accomplished. It is no wonder that sexual reproduction has become the evolutionary technique of choice.

Genetic algorithms have recently become an important research topic for several reasons. They provide a tool that offers an efficient, effective means of searching through a vast array of possibilities. This means that, when combined with other AI techniques, they can be powerful indeed. Remember that one of the fundamental AI problems is how to efficiently search through a large tree of possible solutions to a given problem. Combining genetic algorithms with AI's use of heuristics ("rules of thumb") now appears to offer immensely powerful methods of implementing such searches.

But what do evolution and genetic algorithms have to do with building an android? The fact of the matter is that many androids are likely to be wholly or partially self-reproducing. At the very least, they must be able to detect and diagnose problems within their systems, and more than likely they will have to be able to do at least simple repairs on themselves as well. Why must this be so? Many androids, particularly those used for space exploration, are likely to be operating in environments that are remote or hostile to people. If an android is sent to land on and explore the surface of Mars, it cannot just pop back to the local repair shop if a sensor fails. Instead, such systems will have to be as nearly self-sustaining as possible if we are to achieve the maximum utility from them. Let's consider just one example, drawn from real life.

Christopher Langton of Los Alamos National Laboratory has sketched a fascinating scenario.* NASA is currently considering (among other proposals) the establishment of a permanent base on the Moon. The idea is that the Moon may have access to a number of minerals useful here on Earth. In addition, low lunar gravity may make the Moon a good place for the manufacture of some kinds of products or drugs. At the very least, it would be much easier, cheaper, and more energy-saving to move building supplies from the Moon's surface to Earth orbit—as might be needed, for example in the construction of Space Station Freedom—than from the surface of the Earth because of the Moon's weaker gravity. Now the problem is how to actually build such a moonbase. Suppose we develop an android mining colony (or factory) that we ship to the Moon. The purpose of the androids would be to build their own moonbase, make it operational, and then begin shipping the resulting products to the Earth. These androids would be much cheaper and easier to sustain for long periods in the harsh lunar environment than people—although it is probable that astronauts might be used to assist in the preliminary building phases. But androids don't eat food, require water, or need an oxygen atmosphere, all of which would be exceedingly expensive to maintain on the lunar surface; furthermore, they can be made vastly more resistant to cosmic radiation than people. It is likely to be much cheaper to use them as the primary lunar workforce.

With any mechanical device, there is the likelihood of breakdowns and accidents. If the station is to be nearly or wholly automated, it behooves us to build the androids so that they can maintain themselves. But Christopher Langton's lunar colonization scenario takes this a step further. As we extend our reach into space, it is likely that we will need still more lunar products and building materials, much more than we could produce from a single factory or station. So some of the initial station's output might be diverted to building a second station at some suitable location: the moonbase would be self-reproducing, in other words. This scenario envisions a population of self-reproducing, self-sustaining, robotic mining colonies that expand their population in concert with humanity's increasing need for resources from the Moon.

When reproducing any complex system, errors are bound to creep in, even if—especially if?—the reproduction is carried out partially or wholly by people instead of solely by the androids themselves. (Many examples exist in the plant and animal kingdoms, by the way, of one

*See *Artificial Life,* a volume filled with a number of intriguing glimpses of Artificial Life research. The book contains the proceedings from a workshop held at Los Alamos in September 1987.

species depending on another to complete its own reproductive process.) Some of these errors—like the random genetic mutations that occur in animals—are likely to be meaningless: An android might end up painted black instead of white, or perhaps not painted at all, for example. But others are likely to occur during the process of downloading the control instructions to the android. Such changes, if left unchecked and combined with the power of genetic algorithms and changing environmental pressures—not to mention further human technological development over time—may cause these android colonies to evolve. As a result, the characteristics and behaviors of androids a few dozen generations removed from the initial landing party are likely to be vastly different from those of the original lunar settlers. Because of the evolutionary imperatives, these changes will almost certainly cause the newer generations to be far more suited to their harsh environment than the original models.

Does all this sound a bit farfetched? Well, perhaps it is. Nevertheless, it seems clear that, whatever the initial working android's capabilities, androids ten or twenty or fifty years later will be vastly advanced. And that will be true for androids on Earth as well as in space. Given the unremitting advance of computational technology, it will be astonishing indeed if the evolution of androids does not occur at a blindingly fast speed. Moreover, not all the improvements and adaptations that occur are likely to be enhancements planned by human engineers. Evolutionary effects will almost certainly appear—and in ways we cannot predict.

One side effect of evolutionary processes is the notion of preadaptation. This means that a system that evolves within an animal (or android) to solve one problem may prove to be a good solution for another, very different problem that the animal encounters later. For example, it has been speculated by David Stork of Stanford University and others that human language capabilities are an example of a preadaptive capability; that is, that the structure of the brain that seems so suited to learning and understanding language may well have originally developed for some other purpose entirely. Preadaptation may also help to explain the effectiveness of evolutionary forces. Suppose a pond-living fish species faces a changing environment in which the pond is drying up. Because of some obscure genetic accident, this particular species of fish has developed particularly strong fins, perhaps as a result of species-specific mating rituals. The strong fins mean that this species is partially preadapted to moving on land—that is, it can clumsily use its strong fins to shove itself about on dry (or at least muddy) land. As the pond continues to dry up, the species with strong fins has a survival advantage over those others that do not, and evolutionary selection for primitive legs proceeds.

Such cases may seem—with 20–20 hindsight—to follow some kind of grand plan, but it is not necessary for an overall plan to exist for evolution to take such apparently fortuitous steps. Among the species living in that pond may be any number of other mating-adaptations that have nothing to do with strong fins or living-on-land adaptations. Perhaps one species uses bright coloration to select for mates, another has a very powerful tail, and a third has a large, heavy mouth of teeth that it shows off in a grinning aquatic dance. None of these capabilities is likely to be of the slightest use in moving about on land. The fact is that natural evolutionary forces tend to develop a remarkable diversity among and within species, and that diversity is what enables preadaptive developments to occur. Since future environmental conditions cannot be foreseen, spawning a broad variety of characteristics and abilities is the only certain way of ensuring that at least some of the species can survive no matter what changes occur. When ecological change strikes, the adaptations that can be suitably applied to the new conditions survive, while those that might be nicely capable in the case of a different ecological change (but that are not so capable under these specific new circumstances) do not survive.

Researchers have already created artificial beings that exhibit evolutionary traits. There is a burgeoning field called "artificial life," or AL, that explores the realm of systems that breed, interact, and evolve. In some respects, AL is more closely attuned to the study of neural networks than it is to AI, however. AI is a very *computational* approach to the problem of intelligence. AI techniques are always reducible to an algorithm—a cookbook sequence of specific steps that are taken to solve a particular problem. In contrast, AL, like the field of neural networks, tries to model systems that actually exist in biological systems. In the case of neural networks, the model is the brain—usually the cerebral cortex of a higher mammal. In AL studies, the model is genetics. The three basic aspects of a genetic algorithm—reproduction, crossover, and mutation—are all events that actually occur in animals as they reproduce. Just as neural networks are generally most profitable when they are built on reasonable models of actual brain structures, so too are AL studies more profitable when natural principles of evolution and genetics are followed. It is clear, in fact, that Mother Nature still has much to teach us about intelligence and evolution.

As can be seen from NASA's consideration of self-replicating lunar mining colonies, the field of AL is not without its practical aspects. Interestingly, a number of systems have been developed that, while only existing as computer programs, seem to actually have a life of their own. These simulations are simple beings, not much more complex than Braitenberg's cars, that run independently, interact with each other and their computer environment, feed, breed, evolve, and generally exhibit all the behavioral characteristics of living organisms. Are they alive? Who is to say they are not?

Certainly some computer programs exhibit tendencies that are very difficult not to ascribe as lifelike. Computer "viruses" have received a great deal of publicity over the last few years, and with good reason. These are programs that "infect" computer operating systems, reproduce themselves, and then transmit themselves to other computers that may be in contact with the infected system. Most often, viruses are transmitted by users who do not follow "safe computing" practices. Sometimes these viruses are nothing but a mild annoyance, but some very virulent and destructive versions damage information stored in the computer's memory and disks. Such computer viruses can certainly pass nearly any reasonable test for life imaginable—except that they do not carry their own DNA. Instead, their "genetic code" consists of a sequence of computer instructions that can reproduce itself and otherwise perform its life functions. It should come as no surprise at all to realize that discussions of computer viruses—and their counterparts, "worms" and "Trojan horses"—are replete with images culled directly from biology and medicine. Computer viruses may actually be the very first artificial systems that can be truly called "alive."

We should consider, by the way, whether it is fair to call something "alive" that can only exist in the specialized environment of a computer operating system. The fact is that *all* lifeforms demand a specialized environment of one sort or another. Some creatures—including human beings—are fortunate in that their necessary environment is very broadly defined. Others, such as the giant panda of China, hover on the brink of extinction largely because their acceptable environment is too narrow. It is also completely unfair to restrict the definition of life to only those environments perceived to be "natural," as in created by nature rather than created by mankind. In the first place, can anything created by a child of nature be unnatural? If that is the case, then the environment generated when a beaver builds a dam and floods a stream also must be considered unnatural.

But there is another, more telling, argument as well: Life has a history on Earth of filling every possible environmental niche. Even in the coldest, most barren reaches of the Antarctic desert, life exists in the form of tiny plants that live on and within the rocks. A more unpromising environment for life can hardly be imagined, since it is in near-total darkness for half the year, and suffers extremes of climate that are almost unimaginably fierce. Yet there, too, life exists. Life can be found at the top of mountains and in the depths of the ocean. Life even exists in volcanic steam-vents underneath the sea, in places where no free oxygen is available. Is it not at least possible that in inhabiting the inner reaches of a computer's operating system, computer viruses and their cousins are merely filling in a new environmental niche that never existed before? Granted, the environment is the creation of mankind, but is it not also possible that the tenacity of life makes such creatures as computer viruses inevitable? In that case, the term "artificial life" may truly be a misnomer; the more accurate term might be merely "life."

One of the greatest proponents of AL is Hans Moravec of Carnegie Mellon University. Professor Moravec outlines a possible future of intelligent beings in his book *Mind Children: The Future of Robot and Human Intelligence*. His thesis is that within the next century, human beings will be supplanted by the children of our own minds—androids, in other words. Robotic intelligence and evolution, according to him, are likely to replace DNA-based intelligence on this planet within as little as a hundred years. This seems an extreme position, but it certainly has at least a grain of truth in it. Androids, once built, are almost certain to have some evolutionary capability within them, even if only from mutation and copying errors. If two or more androids are used to create the following generation—android sexual reproduction, in other words—it can be expected that their evolutionary development will proceed extremely rapidly. And we should not forget that mankind is an inveterate tinkerer, so that just because the first version of an android is built does not mean that its creators stop their efforts to enhance and extend its capabilities. The continuing development of androids by human beings (who will certainly control the major developmental thrust of androids initially), combined with their own evolutionary tendencies, may well cause the development of androids to exceed even the dizzying speed of our own evolutionary path.

It has taken a million years (give or take a few) for mankind to change from a small band of wandering ape-beings to the indisputable masters of this planet. By far the majority of that development has been in the last 5000 years. We stand at the crossroads of creating another intelligent set of beings that may eventually exceed our own enormous potential. These "mind children" are certainly the products of our minds and hearts, if not our flesh and blood. Will they be alive, or will they be mere machines? A narrow, parochial perspective may insist that life can only exist when it is controlled by DNA merged into genes. This may prove to be one of the greatest errors of judgment humanity has ever made, for while we treat these mechanical children as machines—while our backs are turned and we are looking the other way, secure in our arrogance that we are masters of our world—as all this is happening, they may be evolving into our successors.

≧ 13 ≦

Redefining the Measure of Mankind

More than at any time in history mankind faces a crossroads.
One path leads to despair and utter hopelessness, the other to total extinction.
Let us pray that we have the wisdom to choose correctly.

Woody Allen

Someday in the not-so-distant future we will awaken to a world that is radically different from the one we know now. We will be sharing our world with the products of our own imagination, androids. What will life be like then? What will happen to our society when we share it with people-like machines? And how soon will all this happen?

To answer the last question first, I believe that the first working android will be built shortly after the turn of the century. Within twenty years, we will have the knowledge and ability to build androids that look and act in a way most people will consider "human-like." Even within a single decade, before or shortly after the year 2000, androids will begin to be a part of everyday life in a way they are not now.

Today's robots are limited mostly to very specific assembly-line tasks or highly specialized environments few people ever encounter. By the year 2000 or shortly thereafter, I believe that most people will have had some interaction with a robot, and many will deal with them on a regular basis. The Japanese have already developed a working robot that can cook and serve microwaved meals; they plan to house it in a mannequin form and use it as a fast-food restaurant "employee." The manufacturer predicts that such robots—or improved versions that can also operate a grill—will be in widespread use in the fast-food industry before the turn of the century. So your future meals at McDonald's or Burger King or Wendy's may be filled by a robotic cook. It is even very possible that the mythic housemaid's robot will be commercially available by then, although it is likely to still be a luxury item beyond the means of most households. There will certainly be android space travelers exploring—or on their way to explore—Mars and the Moon by the turn of the century. If not built by the United States, the Japanese or the Europeans will ensure that humanity's mechanical delegates take the next steps into space.

This prediction may be shocking to many, particularly in light of the enormous distance we have yet to go in developing the necessary systems and subsystems involved in such an undertaking. As we have seen, we have—or are about to have—excellent technologies that can provide memory, learning, training, speech, and vision systems. Each of these areas is either sufficiently advanced today to build the necessary systems for at least a primitive android, or we can expect them to be so within three to five years. Other areas are less well developed, but even these research areas are likely to see advances that will lift them to the minimum necessary levels within just a few years. While the first android systems may not be as smooth and debonair as we might like, it seems evident that they will be built, and quickly too.

This does not mean to imply that all the problems have been solved, nor that all remaining problems are even close to a solution today. As we have seen, many of these are going to be difficult to solve in the available time. Instead of glossing over these difficulties, my prediction is actually an assertion of faith in the ingenuity and talents of the researchers involved. Looking back at the progress that has been made over the last ten or twenty years is exceedingly instructive. The technology that is readily available today was practically unimaginable twenty years ago. I have enough faith in this current generation of researchers to believe that they too will make similar strides into the future.

Furthermore, just as the race to land on the Moon resulted in tremendous by-products that have carried over into nearly every aspect of the way we live—ranging from freeze-dried coffee, to improved ceramics and other materials, to better, smaller, and cheaper computers—so, too, will efforts to develop an android result in the development of products and technologies that will spill over into everyone's lives. Robotics technologies will improve everything from video games to artificial limbs that directly interact with a patient's nervous system so that they feel like natural limbs. Improved vision systems will permit far more sophisticated security systems to be developed, as well as cars and trucks that drive themselves—most likely much more safely than the average big-city cab driver. Development of language understanding systems will give us appliances and cars that talk to us, and dictation machines that truly take dictation and convert it directly to letter-perfect text. Improvements in robotic arm control will of course make factory automation more efficient, but also will make gasoline pumps that are more self-service than anything we have seen before. And once we can add common sense to computers, they will be able to perform tasks for us that we can only dream about now.

The real challenge in building an android is not in these mere physical subsystems. No, the true challenge is in constructing the intelligence needed to make an android useful. I have devoted a considerable amount of space in this book to investigating the difficulties involved in making an android with common sense and a personality. These difficulties are significant and important. Yet computer systems and neural networks today are so far advanced beyond what they were ten or twenty years ago that I cannot believe a similar rate of achievement will not be sustained over the next ten to twenty years. Short of a worldwide economic disruption, it is likely that we will greet the new century with a new collection of mechanistic allies.

The most common way in which future predictions err is that the predictor underestimates the rate of advance in the sciences. Practically no one, twenty years ago, could have predicted the power, flexibility, and ease of use of today's personal computers. The cars we drive now may have sophisticated computers built into them and "heads-up" displays that cast the instrument panel onto the windshield so the driver never has to look down. Only a few years ago the heads-up display technology was a hot new innovation in the most advanced fighter jets; yet today it is available in cars anyone can buy. We are manipulating genes and attacking illnesses that were considered hopeless only a few years ago. Practically none of this could have been predicted with assurance in the early 1970s.

Even more dramatic, however, are some new technological developments being made today. Researchers at NTT Labs (the Japanese equivalent of America's AT&T) have developed a neural network system that quite literally reads a person's mind. The system receives inputs from about a dozen sensors attached to the skull, just as an EEG recorder does. Then the user literally *thinks* the words "up," "down," "left," or "right" and the neural network translates the changes in EEG patterns into controlling commands for a joystick. A similar device in the form of a headband has been produced by other researchers at Stanford University. Interestingly, not only are users such as the U.S. Air Force interested in this technology for use in fighter plane controls—remember the movie *Firefox* in which Clint

Eastwood controlled the weapons system of a new Soviet plane by thinking the appropriate commands in Russian?—but one of the most eager potential customers is a major video game manufacturer. Such "science fiction" technology may end up in American homes by way of a special video game helmet or headband that reads the user's mind.

What will the development of such "telepathic controllers" do to our world? The obvious possibilities are in the control of complex systems, such as airplanes and other machinery. Video games are similarly obvious, as are systems to help the handicapped—imagine a system that could control a robot arm for a quadriplegic, merely by having the person think about what the arm should do. But more astounding possibilities are also likely, many of which we cannot even envision. What about a "telepathic" house, in which all the systems and appliances responded to the thought commands relayed by a simple headpiece, similar to a set of earphones? Or a nursing station that was similarly equipped? Or a car that responds only to the thoughts of its rightful owner? Or a telepathic "realie" that allows viewers to control the action in a virtual-reality environment so that experiences there are tailored to the individual's tastes and preferences? It would surprise me greatly if such "telepathic" controllers were not readily available before or soon after the year 2000.

Companies have already arisen that specialize in developing applications in "virtual reality." These companies produce devices—headbands, gloves, even full bodysuits—that a user wears or otherwise manipulates. The natural gestures of the user are interpreted by high-powered supercomputers and translated into changes in a simulated environment. A simple version of this is used in aircraft flight simulators in which changes to the simulated plane's controls cause the pilot to see changes in the instrument readings and exterior views through the simulator's "windows." More complex versions are being developed for use for such varying applications as interactive games, walk-throughs of buildings that have not yet been built, and truly interactive teleconferencing. Just as the cave paintings of our neolithic ancestors may have allowed hunters to anticipate and learn about the hunt, such virtual realities will provide each of us with a way to experience things that would otherwise be beyond our ken.

Yet another reason future predictions are often wrong is that social changes are nearly impossible to predict, and society determines which technologies are developed and which are ignored. In 1960 John Kennedy issued challenges to the young people of America, and the result was both humanity's first step on the Moon and an outpouring of aid to underdeveloped countries through the Peace Corps. If Richard Nixon had won that Presidential election, would there have been a U.S. Moon landing in 1969? I suspect not. A few thousand people, by voting for Kennedy in the extremely close presidential elections that year, literally changed the course of world history. Was that predictable? Probably not, and certainly not from a decade or more before. Could any prognosticator have predicted the AIDS epidemic and its resulting impact on social and sexual mores? Or the crumbling of the communist bloc in Eastern Europe in late 1989? Or the invention of the transistor? Or television's impact on the family and politics?

The point is that seemingly small events can cause dramatic changes in the history of the world. Yet it really is the desires of society as a whole that determine the direction of technological development. For example, in the 1970s there were a large number of grandiose and complex plans for public transportation systems in various cities in the United States. These largely fell into dust, mainly because the American people don't much like public transportation. We are a society that prefers private automobile transportation, so it is the technology of road-building that is most developed here. In Europe and Japan public transportation is far more acceptable, and therefore much more advanced. Society seems willing to pay any price for a technology that is deemed desirable and worthwhile—and little or nothing for one that is unpopular, or otherwise unwanted. In the 1960s the American people *wanted* to put a man on the Moon; it was a necessary achievement to maintain our image of ourselves as a people. As a result, and in spite of increasing grumblings, the price was paid and our egos soothed. Now, the race for space takes a backseat to other problems—but if it again became a national priority, we can be assured that the resources would be found to reach the desired goals.

The result of all this is that a predictor of technological advances must also be a predictor of sociological trends. He or she must try to anticipate what social pressures will be brought to bear on scientists, and, from that, determine where resources will be allocated. This is an undertaking that is fraught with the likelihood of error and inaccuracy.

There is one other difficulty in predicting the rate of development of new technology, and that is the availability of necessary support systems for that development. For example, neural networks were initially proposed and developed in the late 1950s and early 1960s. At that time, there were no really powerful computer systems, no such thing as a graphical interface to show researchers what was going on in the network, and even high level computer languages such as Fortran and Lisp were merely in their first, primitive stages. On top of all this, no one had yet developed a learning procedure that could handle multilevel networks, so neural network achievements were limited to simple problems. The result of this combination of factors was that this technology fell into disrepute for nearly two decades, until the early 1980s. During that time, computer hardware and software developed to the point where really good simulations of complex networks could be performed and tested. Thus today neural network development is proceeding at an ever-accelerating rate, in large part because the tools and support technology needed for its proper development is now in place, which was not the case twenty years ago.

The future prognosticator must, therefore, take into account the development of these necessary support technologies. For example, in the case of an android, I must assume that tactile and other sensors will be produced as needed, and that computer hardware will also continue to develop in power, speed, and miniaturization. Of all the factors that must go into a prediction of the future, this is the safest. Computing technology has advanced by a fairly constant rate of a 1000-fold improvement every twenty years, and there are no signs that this is slowing down. Thus, the computers of the early twenty-first century can be expected to be a thousand times cheaper-smarter-faster-smaller than the most advanced systems today—and that is very fast indeed. And with the current rate of advance in neural networks and other AI technologies, they are likely to match that development rate, given adequate resources and effort.

A year or so ago before this book was written, I wrote a book with coauthor Charles Butler called *Naturally Intelligent Systems,* in which we compared the then-current state of neural network technology to a fragile seedling, struggling to survive. Now that analogy seems no longer apt. The fragile seedling has become a strong young sapling, filled with vigor and energy and growing rapidly. Similar changes and advances are being experienced in robotics, sensor development, machine vision, and artificial intelligence. Such dramatic growth in science and technology is bound to lead to dramatic changes in the way we live and work.

While interest in these technologies in the United States is still expanding rapidly, the fact is that *both* Europe and Japan are each spending an order of magnitude more—ten times, in other words—what we are in the development of these research areas. In most key technologies, the United States spends far less than our world competitors in primary research and development. The Japanese already have a near-lock on much of the semiconductor industry. We are not yet losing the technology race, but our once-great lead has been whittled to a bare edge—if it exists at all. So the housemaid's robot that is all the rage in 2010 will most likely be made by Sony or

Mitsubishi or Panasonic, unless dramatic changes are made in the U.S. approach to research and development.

The next question, of course, concerns what social changes will result because of the development of robot technology. Scientists and engineers have a strong inclination to list vast amounts of benefits for society and mankind as a whole. Possibilities for these benefits include improved care for the aged and the very young, relief from household drudgery and other unpleasant or dangerous tasks, improved productivity, and many others. Certainly some of these benefits will come true. For example, I suspect that initially most of the impact of intelligent androids will occur in the workplace, rather than at home. Until a housemaid robot costs as little as, say, a quality automobile, not too many households will be able to afford one. Once the price gets down that low, however, the broad middle class will probably find that a robot to keep house, like today's VCR, portable telephone, and video game, is something that they simply cannot live without.

At work it is likely to be a very different story indeed. There are many "drudgery" jobs that a competent android should be able to accomplish with ease. Possible "android-jobs" include: waiters, store clerks, manufacturing assembly-line positions, security guards, farm workers, bank tellers, cab and bus drivers, house-cleaners, maids, nursing aides, nannies, cooks, teachers aides, and dozens of others, including the fast-food restaurant workers mentioned before. Notice that many (though not all) of these positions require few skills and/or earn relatively low pay. If, for a modest capital investment, a company can obtain an android worker that takes no breaks, never is sick or on vacation, never leaves for another company or position, needs no fringe benefits, and works honestly and reliably, then even a high initial purchase price could well make solid financial sense. On the other hand, this could result in real hardship for those humans who lack the technical skills to perform more demanding and creative jobs that androids cannot yet do. There may be far fewer positions available for those people who lack a high school diploma—or perhaps even a college degree—and as a result, unemployment may rise and social services could be strained.

There could be a bright side to such trends in this country, however. It is predicted that in the first half of the twenty-first century, the population of the United States will be growing significantly older. In fact, many fear that as the Baby Boom generation reaches and passes retirement age there will not be enough working adults to support them in their old age. The addition of androids to the workforce, particularly in the low-paying service industry, may help take up the slack from the demographic losses in the population at large.

The answer to the question of whether android workers will be a blessing or a curse to society will likely depend strongly on each individual's position in the workforce. Those on the upper end of the economic scale, and who hold positions that are not likely to be exposed to android competition, will probably consider them to be a boon; those whose jobs and livelihoods are directly threatened will certainly take a darker, more negative view. Which view will dominate is a question that only time can answer.

Given the possibility of (human) worker-revolt as androids begin to compete directly with human workers, what incentive exists for us to build them at all? Why would we develop devices that may prove to be disruptive in the workplace and threaten the job security of human workers? There will be as many answers to this question as there are people who work on android development projects, but I can present a few of the more prevalent reasons here. First, some people will build androids because of the sheer intellectual challenge in creating a device so complex. This is the most difficult problem we have ever tried to solve—much more so than merely sending a man to the Moon—and simple intellectual curiosity and determination will drive many researchers. Others will build androids because they hope to improve the world by providing competent, low-cost (relatively) workers for difficult or dangerous jobs. Some will build them because they think money can be made by doing so—and undoubtedly they will be correct. A few will construct androids because they will be easier to manage than people, less independent and demanding of rights and privileges than a human being placed in the same circumstances. And some will build them just because it's their job to do so.

As time passes, and robot generation succeeds robot generation, their physical appearances will change dramatically. Depending on their individual destinies, each android type is likely to have a specialized form. Some may have multiple hands or legs; others may have eyes on all sides—nannies, like parents, really need eyes in the backs of their heads, for instance. Some will look more like a refrigerator than a person, and others may appear more like an insect. No matter what their physical form, however, the brightest of these, the ones that can truly be called androids, will be those that are destined for general operation in a wide variety of situations.

These general-purpose robots will have to be smarter than those devoted to individual specific tasks in order to cope with a changing set of problems and environments; they are likely, in fact, to be the geniuses of the robot world. Such robots must have a general-purpose form as well, one that may not be optimal for each individual task, but one suitable for many different roles. In nature, the most adaptable animal on Earth is *Homo sapiens,* and our form has the general characteristics needed for such adaptability. We walk erect so we can see more of our environment; we have a couple of manipulating hands and arms; we have a head at the top of the body, with a well-protected braincase and most sensing organs located there as well.

There is a logic to why we are put together the way we are, and there is no reason not to take advantage of that logic when we construct a general-purpose robot. There would be little point in locating eyes in the feet, for example—we couldn't even see over a curb in that case. By locating sensory organs such as eyes and ears in the head at the top of the body, we achieve the maximum range of detection in these senses. Yes, we may give an android an extra hand or two, or improve the design in other ways, but the basic body shape will be a recognizable copy of a human being, and thus will merit the title "android."

There is little doubt in my mind that the smartest robots we build will look reasonably human—our egos, if nothing else, will assure that this is the case. But might they not look *too* human? Might they bear such a close resemblance that we cannot tell them apart from us? This is unlikely as long as we are talking only about mechanically constructed devices. But it is not necessarily true that androids must be constructed mechanically—we might grow them instead.

Biotechnology and genetics are growing as fast or faster even than computer technology. Already a massive, decades-long Human Genome Project is underway that has the goal of completely deciphering the entire genetic code of our species. We are learning how to clone animals, and already have developed microorganisms and even laboratory animals that carry a specific genetic heritage and can accomplish specific tasks, ranging from cleaning up after oil spills to manufacturing desirable drugs. The technology that today manufactures a specialized organism to produce interferon may tomorrow manufacture a device that looks utterly human—but carries an artificially crafted brain.

Legal precedents have been set in these cases: patents have already been issued on several genetically engineered organisms, including a laboratory mouse with a specific genetic heritage. Such precedents will certainly have tremendous long-range implications for ourselves and the beings we construct.

Biological androids, should they ever be built, would have many of the same frailties of a person, but might be manufactured to meet physical, emotional, or intellectual specifications that may be difficult to locate in the general human population. In addition, such beings may offer a "sure thing" in terms of the end result—something that no parent in the ordinary sense can ever be granted. Would you like your son to be an exceptional pianist? Just go down to the bio-android store and select the hair and eye color you'd prefer. Do you want little Susie to be gorgeous and a brilliant scientist? Make your selection and pay your money. The results are perfectly guaranteed, because the devices are built to your specifications, with none of the risk involved in letting your genes randomly merge with those of your partner.*

But even if androids are purely mechanical (as seems most likely to me), they will still take advantage of the general-purpose shape that has served us so well. Their mechanical forms, however, will make them stronger, more durable, and less susceptible to disease and injury than human beings. They may also be able to tolerate an even broader range of environments than people, such as the cosmic radiation found in space or on other planets. This is likely to give them a strong competitive advantage over similarly qualified (but less durable) humans for many jobs. And when combined with their other characteristics such as reliability, and low (or no) fringe benefits and salary, androids may well take over many positions in the workforce.

These issues do not strike at the heart of the impact of androids on society, or how they will change us as a people. The widespread use of intelligent androids is likely to be the most important development in history, and perhaps in the whole evolutionary development of our species. Their effect on us and on our society will be profound and, possibly, devastating. While many of the side effects may be beneficial—improved care for the aged and the very young, for example—the negative side effects from this technology may well outweigh the benefits.

If you considered the arguments presented in Chapter 12, you know that I believe androids will (sooner or later) have the quality we describe as mind. Mind will emerge from the complex interaction of the independent subsystems that make up an intelligent android. We will not need to program it in; we will not have to explicitly build it. We probably won't be able to understand it any better than we understand our own minds—and maybe not as well. But when the android

*The guarantee only applies to those characteristics that are determined solely or primarily by genetics. The dividing line between effects dictated by genetic heritage and those caused by environmental influence is fuzzy and highly disputed. The current trend is to assign greater and greater importance to genetic heritage, with the result that many people may soon find themselves discriminated against in insurance and the workplace for genes they carry that may someday make them susceptible to diseases or other problems. Genetically engineered offspring guaranteed to be free of such failings are likely to become much more attractive should this trend continue to grow.

becomes sufficiently complex and intelligent, self-awareness and mind must inevitably follow. Mind is not likely to be present in the first android, but eventually it will come, and we may not know exactly when that fuzzy line between mind and not-mind is crossed.

On an individual basis, people may have a hard time adjusting to these ever-smarter androids, because while the first ones will certainly be "mere" machines, somewhere in the twenty to fifty years after their initial widespread introduction androids will take on the characteristics of mind and personality. The social rules and laws that will govern the first androids will not be adequate to govern these later, self-aware systems. And that is when the trouble will begin.

You may wonder why I am so sure that androids will eventually be self-aware. I can explain this belief best in the form of a simple thought experiment. Suppose that the first "intelligent" android is, at best, nothing more than a low-grade moron. In fact, suppose that it has an IQ-equivalent of 10, or even 1. This is clearly not a very bright device at all, and one that few people would have any trouble calling a "mere" machine. In fact, it is probably too dumb to even be considered intelligent.* Nevertheless, as I mentioned before, computer technology is racing ahead at the steady rate of a 1000-fold increase in performance every twenty years, and shows no sign of slowing. It is inconceivable that improvements in computational technology will not quickly find their way into the design of current androids. Thus, a mere twenty or so years after that first moronic android, we should expect its successors to be 1000 times as fast, smart, and capable. So the android "Adam" with an IQ of 1 is likely to be succeeded within a single human generation by "Adam X" with an IQ of 1000. (The IQ of the "average" human being is 100 ± 10 or so IQ points.) Even if computer technology development slows drastically, to, say, a 100-fold improvement every twenty years, that still means that the android constructed a mere single human generation after the original one will have a measure of intelligence very comparable to that of the average human being.

If we truly build an intelligent android that can interact with people, experience pain and pleasure, make decisions, operate independently on the job—or at least as independently as a human being—then is this invention still a machine? Or is it really another

*A strong argument could be made that existing robots have already exceeded IQs of 1; for example, WABOT, the Japanese keyboard-playing robot can be argued to be more capable than an IQ of 1 would indicate. Even Rodney Brooks's "spider" robots from Chapter 4 seem more capable than this simple criterion. On that basis, I could argue that the android "Adam" has already been built and the twenty-year countdown to an android of average human intelligence is well underway.

being like ourselves, albeit with a different body chemistry, but still a being to be respected as having self-worth? Have we created a sophisticated machine, or have we invented a new species of being, one that can become our ally—or our enemy?

The questions and problems such a development raises are many and deeply disturbing to a species that has long considered itself the master of all it surveys. Society must consider carefully whether intelligent androids are property or independent beings. What is it that makes a machine a piece of property, and a person not? A mere trick of chemistry—in other words, being based on carbon molecules instead of silicon? It is certainly true that the computer systems we have today have no sense of self-awareness, no recognizable personality, no *mind*.* But will that be true of an android?

At the moment American society cannot even decide when a fetus becomes a human being; it seems unlikely that a possible mechanical intelligence—and an artificial life—would be any less controversial. The legal system today struggles mightily—and sometimes violently—with the issue of what constitutes a human being, always running months, years, even decades behind the latest in scientific advances. When is it acceptable to "turn off" a human being on life support? When is it acceptable to abort a fetus? What constitutes human life? What quality must be present to make the physical being sacrosanct?

Now imagine, if you will, the possibilities for argument, counter-argument, and violent dispute among people who are basically rational and well intentioned, when the concept of *android* life enters this arena. Will any two people agree on which kinds of androids are intelligent beings—and thus deserving of the respect due to any living, intelligent being—and which are not? Will those people who today call themselves "pro-life" take up similar cudgels on behalf of androids? If not, do they not betray themselves as hopelessly bigoted, displaying their (apparent) inability to honor or appreciate any form of life except their own? And those who are "pro-choice" in human affairs: Will they march and protest in favor of the right to choose whether or not to turn off an android?

Will androids become *too* humanlike? Will they become prey to the evils in mankind as well as our virtues? Is an android likely to

*A surprising number of people who work with advanced workstations—Macintoshes, Suns, and the like—may dispute the assertion that their current computers have no recognizable personalities. Fans of the Macintosh in particular frequently ascribe personalities to their systems, as can be seen by the popularity of software packages such as the "Talking Moose" (Baseline Software) and "At Your Service" (Bright Star Technology). Both offer users the chance to interact with "characters" who "live" inside the Macintosh and perform entertaining and useful functions.

develop prejudices, become selfish, or turn violent? We don't now know fully how such emotions work in people, but is it possible for an android to fall in love? To feel jealous? To learn to hate? If this is the case, surely there is little incentive for us to build them, for we are perfectly capable of generating beings with these flaws right in our own species. While androids are not likely to have deep emotions until long after the first one is built, I think it would be a mistake to brush these questions off as irrelevant.

Emotions in people and animals derive at least partially from the necessity of preparing the body for action. Adrenaline surges into the body, readying it for fight or flight responses in emergency situations. Feeling of rage and fear derive directly from such adrenal stimulation. Other feelings such as love and sorrow are less well understood now, but also seen to have a chemical basis, at least in part. In all these cases, however, it is difficult to see what subsystem in an android would provide a similar stimulus as the adrenal system in humans. It is likely that androids will be limited initially to the more sedate emotions: concern, liking, caring, satisfaction, perhaps even sorrow and enjoyment. They may eventually move past these feelings—or then again, they may not.

This emotional stolidity may also add to their reliability in the work place. Certainly it is extremely difficult to envision an android allowing its emotions to interfere with their work, as frequently happens with human beings. It is equally difficult to imagine an android pining away for love, or distraught and grief-stricken from the loss of a friend. Unless there is some evolutionary or performance advantage carried with these profound emotional responses, it seems unlikely that they will ever be part of an android's makeup—in fact, androids may never move past a strong sense of duty and responsibility. Nevertheless, some level of basic emotional response will almost certainly be present, and must be considered when determining how to deal with these beings.

When we construct an android with a built-in sense of pain (for self-protection and self-diagnosis), and at least some minimal collection of emotional responses as that described above, does that mean that we can feel free to ignore those emotions and feelings simply because it is a machine and not a person? If we have a dangerous task that will likely result in the destruction of the android, can we feel free to order the android to carry out the task, dooming it forever? Or must we offer an intelligent android a choice—and what do we do if it refuses to comply?

If we decide that androids are independent beings, how much freedom does that imply? Are they eligible for citizenship? Can they vote? Can they run for office? Can they be managers and supervisors in a factory? (Won't *that* be an interesting tangle, when managers instead of factory workers have to worry about losing their jobs to androids?)

Should it be legal to have a sexual relationship with an android? Is it moral to do so? In fact, should androids be built with any kind of sexual function at all? You may wonder why this would even be an issue, but consider this: Might it not be safer from a hygienic perspective to use androids as sexual surrogates and prostitutes than human beings? At least the possibility exists of disinfecting an android between clients, thus reducing the risk of spreading disease. Could this not be a more socially acceptable way of providing sexual services for hire? Certainly prostitution has existed as long as society—does it not make sense to control the medical consequences by applying some basic principles of hygiene? Humans have consistently manufactured sexual toys of increasing realism and talent; surely someone, somewhere will construct a sexually functional android if only for the titillation value.

But would you want your sister or daughter—or your brother or son—to marry a sexually functional android? Would *you* marry one? (And before you cast this notion aside as being too ridiculous to contemplate, just consider that many people accept same-sex gay marriages as reasonable; marriage today frequently has little to do with procreation. And might not an android step-parent prove to be superior to many human ones?)

And if we are the creators of this race of independent intelligent beings, does that mean that we are on a par with God? What's the difference between breathing life into a lump of clay to make a human being, and constructing an intelligent, self-aware android? Does an android that we create have a soul? Or it is, like Mary Shelley's Frankenstein monster, forever doomed to soullessness? If it has a soul, could an android become a Catholic priest? A Jewish rabbi? A nun? Could an android be a prophet?

If it has no soul that we humans recognize, might androids band together to form their own religion? And, if so, would their God be *Homo sapiens*? Is it not the height of arrogance to assume that we can usurp such a central role in any religion? Or would they even bother with religion at all?

Will the development of intelligent androids lead to riots in the streets as human beings begin to resent the fruits of their own labors? I suspect that such an event is all too probable once their true impact on society is realized. We humans are not noted for our tolerance of beings who are different, and I doubt that a few decades' worth of experience with intelligent androids will be enough to forestall such unrest. The social, legal, ethical, moral, and religious questions raised by the development of a new set of intelligent beings may touch off a social upheaval that makes the 1960s race riots seem mild.

Social upheavals, racism, and other forms of intolerance are usually at least partially triggered by economic competition. How will humans react to competition with the devices we ourselves created? It is difficult to believe that those whose livelihoods and standards of living are threatened by the use of android labor—whether as factory worker, gardener, or white-collar manager—will take such competition calmly. People in general and Americans in particular are quite sensitive to "unfair" competition; an android that needs no employee benefits, and against whose dedication and efforts a human's performance may be compared, is likely to seem exceptionally unfair competition. This could have the dual result of either dehumanizing working conditions for human workers, or even of outlawing android labor altogether.

And what happens to us and our egos when our metallic children outstrip their creators in intelligence? You may have wondered why I seemed so concerned in Chapter 12 with the evolutionary forces acting on androids, when "clearly" android development is in our hands and will not be the result of evolutionary forces. The reason is that as androids become more capable than human beings, they are likely to take over much of their own development and reproduction. After all, would *you* willingly trust the reproduction of your kind to a species who was your own intellectual and physical inferior? Certainly the first ten to fifty years of android development are likely to be guided by humans, but eventually androids themselves will most likely be in charge of their own future. The first android "Adam" may soon be joined by a metallic "Eve." And we may have created our successor—a successor that may exceed our own capabilities in many ways.

None of the questions and issues raised above can be answered today, nor is the scenario of social upheaval I have sketched more than just a single possibility out of hundreds or thousands of other possibilities. But I would like to pose just one more question.

Consider the point I mentioned above that the android worker will never need breaks, never call in sick, never take vacations, never defect to another company, and so on. These are not the attributes of a human worker; they are the attributes of a slave.

A slave is a being that is bound in servitude to another. A slave can make no decisions about his life, not even about his very existence; he is simply bound to obey his master's orders. His life or death is a matter of the whim of his master. A slave is not a person; a slave is a thing. From time immemorial, we as a species have enslaved the animals around us and even others of our own species by considering them as less than human, as mere property. Now we may be preparing to do the same with our own creations.

Suppose we as a society decide that an intelligent android is nothing more than a smarter computer, a machine that is property, with no rights or ability to decide its own fate. In that case, have we not just created a collection of slaves? And in doing so, have we not condemned ourselves? All civilized societies today consider slavery a barbaric, brutish practice that must not be tolerated. Would we not be reinstating an institution that civilized people abhor? Are we as a species so parochial and prejudiced that we cannot see the worth of another kind of being, beings that will be our children in a very real sense, but that differ from us in chemistry and (possibly) in form? And if we cannot cope with the minor differences in form of our own creations, how could we ever cope with the really significant physical differences that are likely should we ever encounter a society from another world or star system?

I believe that society will be placed under strong duress by the invention of an intelligent android. This opportunity presents a challenge to us as humans—presumably civilized humans—to decide whether we choose to react in a civilized fashion, or as slave-holders. Obviously, the human species does not have a very good record in this regard. In our thousands of years on this planet, it is only in the past handful of decades that we have cast aside slavery within our own species; we still abide by the practice with regard to the other species on this planet. We have not yet learned to respect those other biological species, and grant them the same autonomy we have reserved for ourselves. In large measure, I believe this is because of our overweening human ego, combined with other animals' inability to communicate with us on our own terms, in a human language. Androids from the very beginning will have human language as a communications tool, however, and they will be able to tell us what they think and feel and want. I suspect that this single characteristic will make it impossible for us to ignore an android's needs and wants and desires the way we often do with animals. Androids can just keep shouting their needs in our ears until we are forced to listen.

Whether or not we will be sufficiently civilized to take our intelligent creations in stride and deal with them graciously is difficult to predict. Certainly, there will be powerful economic, legal, political, and social pressures exerted to keep our creations as property rather than as partners who share our world. The patents granted for genetically engineered organisms already indicate that we have started down the path of considering them property and not independent beings. But doing so may prove, in the long run, to be disastrous for our society, and possibly for our species. Personally, I am not sure I want to participate in the forcible suppression of the rights of a group of beings who are likely to become my intellectual—and physical—superiors within a few decades. To me, this sounds like a highly dangerous stance to take over the long run.

This may be particularly true if the androids we create absorb more from us than just our intellect. In large part our species has become enormously successful by being, quite frankly, the toughest, meanest, most adaptable animals around. If our android children carry this same tendency, we may easily find ourselves usurped from the position of "King of the Hill." It might behoove us to treat such beings with considerably more care and respect while they are still under our control—the prospect of meaner, smarter, tougher androids coming after us with vengeance on their minds is not a comforting notion to contemplate.

There is one final point I would like to make. Some may think that the way to avoid the conflict and upheaval I have suggested here is merely to refrain from building androids at all. It is true that such a path might avoid the need for humanity to decide how humane we really are. Unfortunately, this is a highly unlikely solution to the dilemma that faces us. Never in our history has mankind *ever* refused to develop a technology that we knew we could exploit. We as a people have built every invention we ever thought of, developed it and tried to make it work. That does not mean that we have always used the invention to its fullest extent once it was complete—the hydrogen bomb and various biological warfare devices are cases in point. Nevertheless, we have always built things once we knew how. I cannot believe that the development of an android will be any different. From whatever variety of motives and methods, we are now standing at the very threshold of knowing how to build an artificial person; I find it extremely unlikely that, once possible, it will not be built by someone, somewhere.

Society as a whole is going to have to learn how to cope with this technology. It will be up to each of us to decide how we should react to it; what limits, if any, are appropriate to apply. The problems involved are knotty and difficult, and will not be solved in a few years or even decades. We must solve them somehow, however, and the way to start is for each of us to begin to consider the implications of such technology. We cannot turn back the clock; neither can we ignore the genesis of intelligent systems that we know—or even just believe—we can construct. To a large extent, humanity is a race of inventors and builders; asking us to not build something new is like asking a bird not to sing—it is counter to our whole nature. For this reason alone, we again must learn to cope with the fruits of our own labors—just as we always have.

The title of this chapter says it all: Intelligent androids will in fact redefine the measure of mankind—but they will redefine the measure of the human species, not that of the beings we create. We will learn much about ourselves in the next decades as we see how we begin to cope with another group of intelligent beings sharing the world with us. It is not likely to be an entirely comfortable time for anyone—but it will certainly be interesting.

There is an ancient Chinese curse that goes, "May you live in interesting times." We have been cursed—and blessed—for we all do live in most interesting times, indeed—and even more interesting times are coming. Whether we are ready to cope or not, the revolution is just around the corner.

Epilogue

The integration of intelligent androids into human society is not likely to be a comfortable process. I have raised a great many questions in this book, yet they are only the tip of the proverbial iceberg. It seems likely that the very existence of intelligent androids will provoke even more questions—questions whose answers we must somehow provide if we are to learn to live with our own creations—and ourselves. It is our generation, and our children's generation, who must decide how to respond to these questions so that we all can learn to live with the consequences of our actions.

In the story that follows, science fiction writer Fredric Brown provides one answer to just one of those questions—and it is an answer that may make all of us uncomfortable. Just remember: The story is only science fiction, not science fact—at least, not yet.

ANSWER

Fredric Brown

Dwar Ev ceremoniously soldered the final connection with gold. The eyes of a dozen television cameras watched him and the sub-ether bore throughout the universe a dozen pictures of what he was doing.

He straightened and nodded to Dwar Reyn, then moved to a position beside the switch that would complete the contact when he threw it. The switch that would connect, all at once, all of the monster computing machines of all the populated planets in the universe—ninety-six billion planets—into the supercircuit that would connect them all into one supercalculator, one cybernetics machine that would combine all the knowledge of all the galaxies.

Dwar Reyn spoke briefly to the watching and listening trillions. Then after a moment's silence he said, "Now, Dwar Ev."

Dwar Ev threw the switch. There was a mighty hum, the surge of power from ninety-six billion planets. Lights flashed and quieted along the miles-long panel.

Dwar Ev stepped back and drew a deep breath. "The honor of asking the first question is yours, Dwar Reyn."

"Thank you," said Dwar Reyn. "It shall be a question which no single cybernetics machine has ever been able to answer."

He turned to face the machine. "Is there a God?"

The mighty voice answered without hesitation, without the clicking of a single relay.

"Yes, *now* there is a God."

Sudden fear flashed on the face of Dwar Ev. He leaped to grab the switch.

A bolt of lightning from the cloudless sky struck him down and fused the switch shut."*

Suggested Reading

The following references provide pointers to further reading on the topics discussed in this book. Sources cited below have been selected for clarity and usefulness for the non-specialist. I would also like to single out Robert Byrne's wonderful collections of quotations, *The 637 Best Things Anybody Ever Said* [Fawcett, 1982] and *The Other 637 Best Things Anyone Ever Said* [Fawcett, 1984]; these books are the source for most of the quotations that appear at the beginning of each chapter of this book. I recommend them highly.

Chapter 1, The Measure of Mankind

"Reality and the Brain: The Beginnings and Endings of the Human Being," by Julius Korein, in *Speculations: The Reality Club*, edited by John Brockman, Prentice-Hall Press, New York, 1990.

The Age of Intelligent Machines, Raymond Kurzweil, MIT Press, Cambridge MA, 1990.

Artificial Intelligence, Patrick Henry Winston, Addison-Wesley, Reading MA, 1984.

The Cognitive Computer, Roger C. Schank with Peter G. Childers, Addison-Wesley, Reading MA, 1984.

Introduction to the Theory of Neural Computation, John Hertz, Anders Krogh, and Richard G. Palmer, Addison-Wesley, Reading MA, 1991.

Neural Network Architectures: An Introduction, Judith Dayhoff, Van Nostrand Reinhold, New York, 1990.

Machines Who Think, Pamela McCorduck, W. H. Freeman and Company, New York, 1979.

Naturally Intelligent Systems, Maureen Caudill and Charles Butler, MIT Press, Cambridge MA, 1990.

Chapter 2, There Is None So Blind

"Applications of Dynamic Monocular Machine Vision" Ernst Dieter Dickmanns and Volker Graefe, in *Machine Vision and Applications*, 1988, 1:241–261.

"Dynamic Monocular Machine Vision" Ernst Dieter Dickmanns and Volker Graefe, in *Machine Vision and Applications*, 1988, 1:223–240.

"Multidimensional Digital Signal Processing: Problems, Progress, and Future Scopes," Nirmal K. Bose, *Proceedings of the IEEE,* special issue on Multidimensional Signal Processing, April, 1990, pp. 590–597.

"An Optical Fourier/Electronic Neurocomputer Automated Inspection System," David E. Glover, *Proceedings of the IEEE International Conference on Neural Networks, July, 1988,* vol. I, pp. 569–576.

Robot Vision, Berthold Klaus Paul Horn, MIT Press/McGraw-Hill, Cambridge MA, 1986.

Visual Cognition and Action, edited by Daniel N. Osherson, Stephen M, Kosslyn, and John M. Hollerbach, MIT Press, Cambridge MA, 1990.

Chapter 3, Taking the First Step

"Invasion of the Insect Robots," David H. Freeman, *Discover,* March, 1991, pp. 42–50.

"Planning," Michael P. Georgeff, *Ann. Rev. Computational Science,* 2:359–400.

"A Robot that Walks; Emergent Behaviors from a Carefully Evolved Network," Rodney A. Brooks, *Neural Computation,* Vol. 1, No. 2, Summer 1989, pp. 253–262.

Introduction to Artificial Intelligence, Eugene Charniak and Drew McDermott, Addison-Wesley, 1985. (See particularly Chapter 9.)

Planning and Understanding, Robert Wilensky, Addison-Wesley, Reading MA, 1983.

Chapter 4, An Android's Reach

"Contact Sensing for Robot Active Touch," by Paolo Dario, in *Robotics Science,* edited by Michael Brady, MIT Press, Cambridge MA, 1989, pp. 138–164. A superb discussion of a fascinating humanlike robot finger; this includes an excellent review of the issues involved in constructing such a finger.

"Development of Neural Network Interfaces for Direct Control of Neuroprostheses," Eric A. Wan, Gregory T. A. Kovacs, Joseph M. Rosen, and Bernard Widrow, in *Proceedings of the International Joint Conference on Neural Networks, January 1990,* edited by Maureen Caudill, Lawrence Erlbaum and Associates, Hillsdale, NJ, 1990, volume 2, pp. 3–21.

"A Hierarchical Model for Voluntary Movement and Its Application to Robotics," Mitsuo Kawato, Yoji Uno, Michiaki Isoe, and Ryoji Suzuki, in *Proceedings of the IEEE First International Conference on Neural Networks,* edited by Maureen Caudill and Charles Butler, IEEE Catalog No. 87TH0191–7, volume 4, pp. 573–582.

"A Six-Degree-of-Freedom Magnetically Levitated Variable Compliance Fine-Motion Wrist: Design, Modeling, and Control," Ralph L. Hollis, S. E. Salcudean, and A. Peter Allan, *IEEE Transactions on Robotics and Automation,* Vol. 7, No. 3, June, 1991, pp. 320–332. In spite of the intimidating title, this article provides a very nice review of current manipulator and robot-arm technology; it also describes the authors' current work in the field.

Neural Networks for Control, edited by W. Thomas Miller, III, Richard S. Sutton, and Paul J. Werbos, MIT Press, Cambridge MA, 1990.

Neurodynamics of Adaptive Sensory Motor-Control, Stephen Grossberg and Michael Kuperstein, Pergamon Press, New York, 1989. A mathematical and difficult volume, but interesting and worthwhile.

A Robot Ping-Pong Player, Russell Anderson, MIT Press, Cambridge MA, 1985.

Robotics Science, edited by Michael Brady, MIT Press, Cambridge MA, 1989.

Chapter 5, Rembering the Past . . .

"Bidirectional Associative Memories," Bart Kosko, *IEEE Transactions on Systems, Man, and Cybernetics*, vol. SMC-L8, pp. 49–60, Jan/Feb 1988.

"Neural Networks and Physical Systems with Emergent Collective Computational Abilities," J. J. Hopfield, *Proceedings of the National Academy of Sciences*, vol. 79, pp. 2554–2558.

"Neurons with Graded Response Have Computational Properties Like Those of Two-State Neurons," J. J. Hopfield, *Proceedings of the National Academy of Sciences*, vol. 81, pp. 3088–3092.

Neurocomputing: Foundations of Research, edited by James A. Anderson and Edward Rosenfeld, MIT Press, Cambridge MA, 1988. For anyone interested in the historical roots of the field of neural networks, or who simply prefers to have access to the classic papers in the field, this book is unbeatable. It reprints key papers, with the added attraction of clear and superbly written introductions that place each paper in perspective.

Parallel Models of Associative Memory, edited by Geoffrey E. Hinton and James A. Anderson, L. Erlbaum and Associates, Hillsdale, New Jersey, 1981.

Chapter 6, . . . As a Lesson for the Future

Adaptive Pattern Recognition and Neural Networks, Yoh-han Pao, Addison-Wesley, Reading MA, 1989.

Introduction to the Theory of Neural Computation, John Hertz, Anders Krogh, and Richard G. Palmer, Addison-Wesley, Reading MA, 1991.

Machine Learning, Ryszard S. Michalski, Jaime G. Carbonell, and Tom M. Mitchell, Tioga Publishing Company, Palo Alto, California, 1983.

Neural Network Primer, Maureen Caudill, Miller-Freeman, San Francisco, 1989.

Neurocomputing, Robert Hecht-Nielsen, Addison-Wesley, Reading MA, 1990.

Parallel Distributed Processing (2 volumes), David Rumelhart, James McClelland, et al., MIT Press, Cambridge MA, 1986.

Understanding Neural Networks: Computer Explorations (2 volumes with software), Maureen Caudill and Charles Butler, MIT Press, Cambridge MA, 1991.

Chapter 7, Discovering the Truth

The Adaptive Brain (2 volumes), edited by Stephen Grossberg, Elsevier Science Publishers, Amsterdam, 1987.

Adaptive Pattern Recognition and Neural Networks, Yoh-han Pao, Addison-Wesley, Reading MA, 1989.

Neural Network PC Tools, edited by Russell C. Eberhardt and Roy V. Dobbins, Academic Press, San Diego CA, 1990.

Self Organization and Associative Memory, Teuvo Kohonen, Springer-Verlag, New York, 1989 (3rd edition).

Chapter 8, It's a Puzzlement

Artificial Intelligence, second edition, Elaine Rich and Kevin Knight, McGraw-Hill, New York, 1991.

Building Expert Systems, Frederick Hayes-Roth, Donald A. Waterman, and Douglas B. Lenat, Addison-Wesley, Reading MA, 1984.

Expert Systems and Fuzzy Systems, Constantin Virgil Negoita, Benjamin-Cummings Publishing Company, Menlo Park, California, 1985.

Introduction to Expert Systems, Peter Jackson, Addison-Wesley, Reading MA, 1986.

Heuristics, Judea Pearl, Addison-Wesley, Reading MA, 1984.

Knowledge Engineering, Steven Tuthill, Tab Books, 1990.

Neural Networks and Fuzzy Systems, Bart Kosko, Prentice-Hall, New York, 1991.

Probabilistic Reasoning in Intelligent Systems, Judea Pearl, Morgan-Kaufman, San Mateo CA, 1988.

Thinking: An Invitation to Cognitive Science, Volume 3, edited by Daniel N. Osherson and Edward E. Smith, MIT Press, Cambridge MA, 1990. See especially Chapter 5 on "Problem Solving," by Keith J. Holyoak, and Chapter 7 on "Rationality" by Stephen P. Stitch.

Thinking, Problem Solving, Cognition, Richard E. Mayer, W. H. Freeman and Company, New York, 1983.

Chapter 9, Speaking in Tongues

Inside Computer Understanding, Roger C. Schank and Christopher K. Riesbeck, L. Erlbaum and Associates, Hillsdale, New Jersey, 1981.

Language: An Invitation to Cognitive Science, Volume 1, edited by Daniel N. Osherson and Howard Lasnik, MIT Press, Cambridge MA, 1990.

Natural Language Processing, Harry Tennant, Petrocelli Books, Princeton, New Jersey, 1981.
Principles of Artificial Intelligence, Nils J. Nilsson, Tioga Publishing Company, Palo Alto, California, 1980.
Scripts, Plans, Goals, and Understanding, Roger Schank, L. Erlbaum and Associates Publishers, Hillsdale, New Jersey, 1977.
Semantic Information Processing, Marvin Minsky, MIT Press, Cambridge MA, 1968.
Strategies for Natural Language Processing, edited by Wendy G. Lehnert and Martin H. Ringle, L. Erlbaum and Associates Publishers, Hillsdale, New Jersey, 1982.

Chapter 10, Sense and Sensibility

"Analysis of Hidden Units in a Layered Network Trained to Classify Sonar Targets," R. Paul Gorman and Terrence J. Sejnowski, *Neural Networks,* Vol. 1, No. 1, pp. 75–89.
"The Two Faces of Creativity," Morris Berman, in *Speculations: The Reality Club,* edited by John Brockman, Prentice-Hall Press, New York, 1990.
"The Mechanics of Creativity," Roger Schank and Christopher Owens in *The Age of Intelligent Machines,* Raymond Kurzweil, MIT Press, Cambridge MA, 1990, pp. 394–397.

Chapter 11, I Think, Therefore I Am—I Think?

"Psychotherapy Systems and Science," Robert Langs, in *Speculations: The Reality Club,* edited by John Brockman, Prentice-Hall Press, New York, 1990.
"Minds, Brains, and Programs," John Searle, *The Behavioral and Brain Sciences,* vol. 3, 1980, pp. 417–424.
"The Uses of Consciousness," Nicholas K. Humphrey, in *Speculations: The Reality Club,* edited by John Brockman, Prentice-Hall Press, New York, 1990.
"The Evolution of Consciousness," Daniel C. Dennett, in *Speculations: The Reality Club,* edited by John Brockman, Prentice-Hall Press, New York, 1990.
"Simulations of Reality," William H. Calvin, in *Speculations: The Reality Club,* edited by John Brockman, Prentice-Hall Press, New York, 1990.

Chapter 12, But Is It Alive?

"Artificial Life," Christopher G. Langton, in *Artificial Life,* edited by Christopher G. Langton, Addison-Wesley, Reading MA, 1989.
"Artificial Organisms: History, Problems, Directions," Richard Laing, in *Arti-*

ficial Life, edited by Christopher G. Langton, Addison-Wesley, Reading MA, 1989.

"Biological and Nanomechanical Systems: Contrasts in Evolutionary Capacity," K. Eric Drexler, in *Artificial Life,* edited by Christopher G. Langton, Addison-Wesley, Reading MA, 1989.

"Human Culture: A Genetic Takeover Underway," Hans Moravec, in *Artificial Life,* edited by Christopher G. Langton, Addison-Wesley, Reading MA, 1989.

"Nanotechnology with Feynman Machines: Scanning Tunneling Engineering and Artificial Life," Conrad Schneiker, in *Artificial Life,* edited by Christopher G. Langton, Addison-Wesley, Reading MA, 1989.

"Preadaptation in Neural Circuits," David G. Stork, Scott Walker, Mark Burns, and Bernie Jackson, in *Proceedings of the International Joint Conference on Neural Networks, January, 1990,* edited by Maureen Caudill, L. Erlbaum and Associates, Hillsdale, New Jersey, 1990, volume 1, pp. 202–205.

Artificial Life II, edited by Christopher G. Langton, Charles Taylor, J. Doyne Farmer, and Steen Rasmussen, Addison-Wesley, Reading MA, 1991.

The Ascent of Mind, William H. Calvin, Bantam, New York, 1990.

The Blind Watchmaker, Richard Dawkins, W. W. Norton & Co., New York, 1987.

Chaos, James Glieck, Viking, New York, 1987.

The Emperor's New Mind, Roger Penrose, Oxford University Press, New York, 1990.

Genetic Algorithms in Search, Optimization, and Machine Learning, David E. Goldberg, Addison-Wesley, Reading MA, 1989.

Mind Children, Hans Moravec, Harvard University Press, Cambridge MA, 1988.

Mind, Machines, and Human Consciousness, Robert L. Nadeau, Contemporary Books, Chicago, 1991.

Mindwaves: Thoughts on Intelligence, Identity, and Consciousness, edited by Colin Blakemore and Susan Greenfield, Basil Blackwell Ltd., Oxford, U.K., 1987.

Neurophilosophy, Patricia Churchland, MIT Press, Cambridge MA, 1988.

Vehicle: Experiments in Synthetic Psychology, Valentino Braitenberg, MIT Press, Cambridge MA, 1984.

Chapter 13, Redefining the Measure of Mankind

"Behind Closed Doors: Unlocking the Mysteries of Human Intelligence," Robert J. Sternberg, in *Speculations: The Reality Club,* edited by John Brockman, Prentice Hall Press, New York, 1990.

"We Aren't Ready for the Robots," Noel Perrin, editorial page of *Wall Street Journal,* February 26, 1986.

Gødel, Escher, Bach: An Eternal Golden Braid, Douglas R. Hofstadter, Vintage Books, 1980.

Index

f indicates a figure reference, *n* indicates a footnote reference